THE STEM CELL DILEMMA

The Scientific Breakthroughs, Ethical Concerns, Political Tensions, and Hope Surrounding Stem Cell Research

2nd Edition

Leo Furcht, MD, and William Hoffman

Arcade Publishing
New York

Arcade Publishing books may be purchased in bulk at special discounts for sales promotion, corporate gifts, fund-raising, or educational purposes. Special editions can also be created to specifications. For details, contact the Special Sales Department, Arcade Publishing, 307 West 36th Street, 11th Floor, New York, NY 10018 or info@skyhorsepublishing.com.

Arcade Publishing® is a registered trademark of Skyhorse Publishing, Inc.®, a Delaware corporation.

Visit our website at www.arcadepub.com.

10 9 8 7 6 5 4 3 2 1

Library of Congress Cataloging-in-Publication Data is available on file.

ISBN: 978-1-61145-352-2

Printed in the United States of America

Genius lives on, all else is mortal.

—Andreas Vesalius
De humani corporis fabrica libri septem
(On the fabric of the human body in seven books)
1543

This really revolutionary revolution is to be achieved, not in the external world, but in the souls and flesh of human beings.

—Aldous Huxley
Foreword, *Brave New World*
1946 edition

I just don't see how we can turn our backs on this.

—Nancy Reagan
2004

CONTENTS

PREFACE

Two years before the turn of the last millennium, a story appeared on the front page of *The New York Times* that sowed the seeds of a dilemma. Written by veteran science reporter Nicholas Wade, the story was headlined "Scientists Cultivate Cells at Root of Human Life." Wade's opening sentence made it clear this was not just another of the research advances that occupy an ever-growing segment of the daily news: "Pushing the frontiers of biology closer to the central mystery of life, scientists have for the first time picked out and cultivated the primordial human cells from which an entire individual is created."

The story surely struck many if not most of its readers in a personal way. It was about possibilities for their own health and that of their family. It was about hope for patients who deal daily and hourly with debilitating diseases. And it was about questions we have not wanted to ask about what it means to be human; about whether the early human embryo has the same moral status that we do or whether it has a lesser moral status or no claim to a moral status at all. The story on the front page of the *Times* that day in November 1998 was merely the first paragraph of the first chapter of a much longer story, a story that has continued down to the present time. It is that story that we set out to tell.

The public policy implications of James Thomson's successful experiment in creating the first human embryonic stem cell line at the University of Wisconsin–Madison remained largely beneath the radar screen for almost three years. Then, one month before the terrorist attacks of 9/11, President George W. Bush addressed the nation from his ranch in Crawford, Texas. The subject of his speech was human embryonic stem cell research. No

American president had ever before addressed the nation like this specifically on the ethics of biomedical research. As a presidential candidate, Bush took the position that taxpayer funds "should not underwrite research that involves the destruction of live human embryos." Bush faced a dilemma: Would he stand firm on his campaign pledge? Or would he allow research to proceed with federal funding?

In his speech, Bush said that embryonic stem cell research "offers both great promise and great peril. So I have decided we must proceed with great care." Private research, he observed, had already yielded more than sixty genetically diverse stem cell lines with the ability to regenerate themselves indefinitely. With that lead-in, Bush set the policy that would determine the federal government's role in the research for the next seven years of his presidency: "I have concluded that we should allow federal funds to be used for research on these existing stem cell lines, where the life and death decision has already been made." Thanks to prior efforts, it appeared, some federal funds would flow into the new research field. News coverage and the public debate that followed over the wisdom of the policy lasted through September 10, 2001, when *BusinessWeek* urged Bush to rethink his position in a commentary headlined: "Stem Cell Science Needs More from Uncle Sam."

The next day, September 11, 2001, the debate came to an abrupt halt. For a time, it disappeared entirely. But over time it came roaring back. By 2005, the debate found its way into the halls of Congress where lawmakers crafted legislation that would have eased Bush's restrictions on federal funding for embryonic stem cell research, legislation that Bush vetoed twice after it had passed both houses of Congress. Bush's restrictions were in fact eased by his successor in the White House, Barack Obama, through an executive order. But that did not end the debate. By late 2010, the future of federal funding for human embryonic stem cell research was in the hands of a federal appeals court. Indeed, there was a real possibility that the U.S. Supreme Court would eventually outlaw federal support for the cutting-edge research field in the country that launched it.

Americans are ambivalent about some things, but the quality of their health care is not one of them. They want the best and are willing to support cutting-edge research with their taxes to find new effective treatments and possible cures. They always have, especially for the past half-century. Our personal experience in the biomedical research field and with community groups, health associations, patient advocacy organizations, and legislators, among others, reminds us always that there is no public appetite to see critical and exciting advances in biomedicine occur someplace else rather than in the United States. As it happens, we also have an abiding interest in how free inquiry, with its roots in the Renaissance, the Scientific Revolution, and the Enlightenment, has improved our lives. It is our experience with the benefits publicly funded science bestows on society and our personal interest in the history of science and medicine that inspired us to write this book—that plus our conviction that we are truly on the verge of something remarkable that will shape the world to come.

Stem cells are nothing new in the clinic. They have been used to treat patients for forty years in the form of bone marrow transplants, for it is the stem cells in donor marrow that rebuild the blood system of the patient receiving the transplant. The first successful bone marrow transplant was accomplished in 1968 by pediatric immunologist Robert Good and his team at the University of Minnesota, the institution where we work. The patient was a four-month-old boy suffering from a deadly immune disease that had already killed his brother. His sister rescued him. Her bone marrow supplied the blood-forming stem cells that replaced her infant brother's diseased cells and restored his immune system to health.

Thirty years later, again at the University of Minnesota, the gender tables were turned: a brother rescued his sister. As you will learn in more detail in chapter 1, six-year-old Molly Nash suffered from Fanconi anemia, a severe blood disease. To save her, her parents produced a number of embryos through in vitro fertilization, one of which became Molly's sibling. Molly's recovery began the day in September 2000 that stem cells from her brother's umbilical cord, which matched her tissue, entered her body.

Like an ever-growing number of people, Molly Nash was saved by stem cells from an umbilical cord, the tether of fetal life. One day five hundred years ago, Leonardo da Vinci held an umbilical cord in his hands and drew it into his anatomical masterpiece *The Fetus in the Womb*. He pondered the mystery of reproduction and development in a room filled with corpses and their contents— organs, vessels, muscles, bone, and limbs. He undertook the exploratory task at a time when such dissections, in the words of his biographer Charles Nicholl, "were beset by taboos and doctrinal doubts." After negotiating the line between curiosity and fear, he resolved his dilemma by venturing into the cave of the unknown to see what he might find. Through his magnificent drawings of what he revealed with his own hands, and because he was convinced that "science comes by observation, not by authority," he lit a flame that has burned brightly in the corridors of free inquiry down to the present day. In observing the umbilical cord, he wrote, "The navel is the gate from which our body is formed by means of the umbilical vein."

Molly Nash's story shows why stem cells are agents of hope for patients and their families. Though stem cells from her brother's "umbilical vein" reformed her bone marrow, the stem cells in umbilical cord blood do not build all the tissues of the body. They do not make hearts, pancreases, livers, kidneys, skin, eyes, bone, and brain. The cells of the early embryo and their successor cells do. They create all the tissues of the body. They build us from when we were visible only through a microscope to what we are today. In late 2007, scientific reports of reprogrammed skin cells that behave like embryonic stem cells created a media frenzy. In the years since, these genetically reprogrammed cells have proven to have capabilities similar to those of embryonic stem cells. They may be able to make all the tissues and organs of the body and possibly to serve as the basis for cell therapies, but that we won't know for some time. One thing we do know, however, is that cells in the early embryo are the architects of development, because they are so versatile.

While many people considered Molly Nash's rescue to be a wonderful story of what modern medicine can do, it was not well received by all. The idea of creating embryos and then selecting one to provide therapy for a sick sibling raised familiar concerns about "designer babies" and new concerns about "savior siblings" as *Newsweek* headlined a story about the Nashes, "A Quick Genetic Test Is a Godsend and a Moral Dilemma." For many people, perhaps most, it was a matter of saving a life and breaking new ground in medicine. For some it was more a matter of destroying embryos and breaking ethical boundaries. It is through politics that your ethics and our ethics and the ethics of the man or woman on the street find their expression in law. Embryo politics are not going away, not even with dramatic research advances using nonembryonic cells. The Japanese scientist who reprogrammed skin cells to function like embryonic stem cells acknowledged the possibility that eggs and sperm could be made using these cells. That would enable same-sex couples to conceive their own genetic child, he told a newspaper. Reproduction technologies, such as those that give many thousands of infertile couples hope of having a baby and restored Molly Nash to health, are here to stay.

From the beginning of the human experience, dreams of re-generation and immortality run like river currents through all cultures. What is different today is our capacity to understand and our growing ability to control the basic unit of life—the cell. Because stem cells in the early embryo direct the development of the organism, understanding that process has enormous implica-tions for medicine and health care. To capture the unparalleled versatility of stem cells, to make "regenerative medicine" a reality, will take a lot of work. It will be necessary to figure out how to direct these cells down the development pathway so that they can be used to repair diseased or damaged tissues. Once differenti-ated into the proper type of cell, they would need to be grown in pure populations and then delivered safely and effectively to the disease or injury site in the body. That would mean for medicine what the moon shot meant for space exploration and what the invention of the transistor meant for electronics. That is why the stakes are so high and why countries, states, provinces, and institu-tions around the world are funneling funds into the new research field. The populations of many advanced industrial countries are aging rapidly. Given the toll that progressive diseases like heart disease, degenerative diseases like Alzheimer's disease, and condi-tions like adult-onset diabetes take on both public and household budgets, the race is on to find more effective treatments and pos-sible cures.

Over the past two centuries, most of the major advances in medicine have taken place in Europe and North America. But such developments as anesthesia, antibiotics, immunization, and transplant surgery are no longer the birthright of the West, if they ever were. Singapore, China, India, South Korea, Taiwan, and other Asian countries are investing heavily in stem cell research, and without heated public debate over the moral status of the human embryo. Across the globe, among states within nations, and even among research institutions, stem cells have become tools of competition.

The power of stem cells and the ability to control what they do could have a darker side. It is no accident that defense agencies are funding stem cell research. They have an obvious interest in fields like wound healing and tissue, nerve, and even limb regeneration. Stem cells fit the bill. But the monumental push in the wake of 9/11 to understand how best to protect the human immune system in the event of a bioweapons attack has helped to turn a page in biomedical science that cannot be turned back. The "push" is backed up by a massive infusion of federal dollars. Stem cells and immune cells are being recruited to assist in the task of detecting the effectiveness of new vaccines without using laboratory animals. The goal is to replicate the human immune system in a laboratory machine called a bioreactor. The day the genius of human immunity is even approximated in a laboratory machine—and make no mistake, we are headed down that road—our collective security is paradoxically both enhanced and potentially compromised. It points up the "dual-use" research dilemma of the life sciences, that biological research with a legitimate scientific or medical purpose could be misused to pose a biological threat to public health or national security, or both.

Looking back on the last millennium of medicine, the *New England Journal of Medicine* editorialized in January 2000, "No one alive in the year 1000 could possibly have imagined what was in store." After sleeping for five hundred years, Western medicine took off, beginning with anatomical exploration of the human body during the Renaissance. The past two centuries have witnessed the sorting out of the role of cells in health, disease, and reproduction. We have discovered how microorganisms like bacteria and viruses cause disease. We have unraveled the mystery of genes in heredity and disease. We have discovered antibiotics and developed vaccines that have transformed public health. We have brought biomedical imaging to the clinic, beginning with X-rays. Today we can scan the process of thought by functional magnetic resonance imaging (MRI). We have witnessed the blossoming of immunology and organ transplantation. In recent decades we have found in nature, and designed in the laboratory, molecules to treat cancer, heart disease, neurological disease, metabolic disease, autoimmune disease, and diabetes.

In 2011, *Time* magazine listed stem cell research among "10 ideas that will change the world: Our best shots for tackling our worst problems, from war and disease to unemployment and deficits." Will stem cells live up to their top billing as the next revolution in medicine? If they do, will patients in the United States be able to go to their local health care provider to receive stem cell treatments? Or will they need to take a flight to some overseas location to receive effective and affordable therapy, as some patients are already doing for certain types of surgery? Will divisive political debates over bioethics determine where research is done and who will fund it? Will the United States be able to retain its homegrown scientific talent as well as foreign students and researchers that are so critical to our leadership in the life sciences? To address these and other challenges we will need to exercise our imagination, individually and collectively, in new ways.

FOREWORD

You know that a field is moving quickly when a second edition of a book is called for only three years after the first. At the same time, many issues have not changed. The underlying technology is still promising, many diseases remain uncured, and the stem cell "dilemma" is still with us as evidenced by the recent U.S. district court case, which prohibited the use of federal funds for not only deriving new human embryonic stem cell lines but even working with existing ones. Although the decision was overturned by a federal appeals court, the issue is likely to remain before the courts for some time. However, the fact that there is active social debate about a technology and the government's role in funding that technology, despite years of congressional funding to date and a growing acceptance among the public, shows that the "stem cell dilemma" still confronts us. With the advent of reprogrammed adult cells over the last few years, many people have said that we no longer need to work with embryonic stem cells and the "ethical" issues have gone away. As this book will show you, such is not the case. The advances in the technology on many fronts have been impressive, but we still have not shown that reprogrammed adult cells, called induced pluripotent stem cells (iPS cells), are the equivalent of embryonic stem cells (ESCs), and we still have not dealt fully with all the ethical implications of the time, assuming that the time comes, when we are able to reprogram any cell to become any other cell.

This book explores such issues and articulates why this "stem cell dilemma" exists and how we can work toward finding some of the answers. Meanwhile, many people and organizations around the world are trying to tackle these questions in both the private and public sectors at the national, state, and local levels. It is an exciting time as new technologies have been developed, as new companies have been created, and as new academic centers have been formed.

One type of organization that has risen to address this challenge is the academic research center. Since stem cell science is not just a new technology confined to the research lab but also has clinical implications, is inherently multidisciplinary, and raises attendant ethical and political questions, universities have realized that their obligations to society as leaders in education, research, and clinical care put them in a unique position to marshal their multiple resources to tackle the problem and seize the opportunity. Such a collection of skills and resources is needed because moving from "bench to bedside" requires multiple domains of knowledge, whether understanding the pathogenesis of a particular disease or knowing how the differentiation and growth processes of one cell type differ from those of another. And these relevant domains of knowledge are not limited to science and clinical care. Universities can draw upon the expertise of the faculties of their schools of law, business, and divinity, in addition to the undergraduate, graduate, and medical school faculties. As a result, the multiple social, political, religious, ethical, and financial issues that surround stem cell research can be explored in depth simultaneously.

In fact, the continuing formation of these centers has led to a movement of cross-center collaboration as consortia are being established, with each center as a local point of integration. It is too early to tell whether there are meaningful economies of scale and scope across centers but the fact is that groups are actively discussing widespread collaboration at scale. This fact alone presages an interesting time ahead as people are figuring out new approaches to the scientific research process and the political discourse surrounding that process.

But universities cannot engage in this work entirely on their own. Partnerships with the commercial sector and alignment with the public and policymakers are needed. And this alignment is highly variable on the national, let alone global, stage. However, one sector that has become much more involved in the last few years is the commercial sector. Many biopharmaceutical companies now have active stem cell research programs—some in cell therapy, but most using cells as tools for drug discovery. The power of stem cells as tools is just being uncovered. The ultimate promise of these tools is to revolutionize the drug discovery and development process. As a result of our ability over the last few years to grow and differentiate embryonic stem cells, reprogram adult cells, and turn one type of adult cell into another, we now have the capability to use stem cells to develop models of human disease using human cells in which to discover and test potential drugs. For the first time, we can grow a human cell of interest from a patient with a given disease and look for a drug that can affect that particular diseased cell type. This is important because drug discovery is a long (fifteen years) and expensive (over $1.2 billion) process for each drug that ultimately makes it to market. Even when a drug makes it to late stage clinical trials, there is a 40 to 50 percent failure rate. Stem cells provide the opportunity to fundamentally change this paradigm, and change the economics of drug discovery, because we can now use human cell–based screening systems to

- understand disease mechanism by observing affected cells as they develop,
- understand the effect of a particular drug on a particular cell type (both diseased and normal), and
- discover how environmental factors can contribute to the origin of diseases.

This will allow for the in vitro study of disease mechanism, therapeutic screening, and toxicology testing all *prior* to studies in people, resulting in drugs that are safer, more effective, and can be brought to market more quickly. Admittedly, human cells in a petri dish do not have the environmental and structural complexity of the human body, but one could argue that it is much more ethical and humane to experiment on human cells in dishes than on human beings—which is essentially our drug discovery process today. As bioengineering techniques and materials continue to become more sophisticated, so too will these in vitro models, in the form of three-dimensional scaffolds and artificial organs. Creating disease-specific cell lines to identify and test drugs will still require validation in accepted animal and human models, but those experiments will be more targeted and safer if this approach is successful.

The recent achievement of reprogramming adult human cells rightly gained widespread attention, with commentary focusing on two items: one, the need to continue with embryonic stem cells; and two, some asserting that they were right all along, that there is no need to work on embryonic cells and that the ethical controversy is over.

On the first point, many articles and commentaries have pointed out the reasons to continue embryonic stem cell work and multiple approaches to reprogramming, including opinions written by the authors of the original studies. Scientists have been careful to call these reprogrammed adult cells "embryonic-like," noting that it is not yet clear that these cells are fully equivalent. In fact, the only way to know that is to have other cells with which to compare them. Recent detailed epigenetic analysis—the study of changes in gene activity that do not involve alterations to the genetic code but can still get passed down to offspring—points to both similarities and differences between reprogrammed adult cells and embryonic stem cells. Knowledge of the one fuels knowledge of the other. As some have noted, the breakthrough in reprogramming adult cells to pluripotency occurred in large part because of what we know about embryonic stem cells. Moreover, there is the issue of trying to figure out how to replace the current reprogramming agents, some of which are cancer-causing genes, with more benign approaches such as using RNA or chemical compounds instead of viruses. Each scientific advance continues to bring new questions as well as new answers.

On the second point, the ethical controversy goes away only if one believes that the early embryo at the blastocyst stage is the full moral equivalent of a person and therefore that derivation of cells from it resulting in the death of that embryo is the equivalent of homicide. It seems clear that although one can grant the early embryo moral value and privilege, it is still not the equivalent of a person. However, as Furcht and Hoffman point out, where the ethical issues continue to confront us is in what happens "going forward." For example, proof that the reprogrammed adult cells in mice were fully functional took several forms including an experiment that turned them into live mice. Primate studies to date have been unable to develop viable organisms from clones, but we can expect the science to advance. Most scientists are clear that they would never do this in humans, but the theoretical potential is there. This raises issues not only about explicitly banning reproductive cloning but also about how to manage stem cell research ethically.

Since a skin cell now represents not just a piece of skin but a potential other cell type, organ, or maybe someday, person, how should we regard it? Does destroying a skin cell destroy potential life, as with an embryo? Stem cell research oversight committees (SCROs or ESCROs) were originally established to look out for the welfare of the embryo. Do those in favor of limiting research to adult stem cells now see these review boards as passé? Or do such boards have even more challenging tasks to think about as cells can become other types of cells, as cells and cell lines are developed that could be useful in a dish but cause cancer in animals, and as such cells are tested in animals and integrated into animal systems? For example, we could develop human neurons with Parkinson's disease from a reprogrammed skin cell and insert them into a mouse brain to study the development of the disease. Questions about what living means, and the boundary between animal and human, will only get more complex. The hard questions are not concerned with "Where did these cells come from?" but "What are we going to do with them?" The good news is that many review boards are already dealing with such questions. And as Furcht and Hoffman remind us, such questions are not inherently different from those we have faced since the time of Leonardo da Vinci each time a major scientific breakthrough was made or a new technology introduced.

These questions are also now facing us in the course of using cells as therapies in people. In the last couple of years, the number of such clinical trials has mushroomed. There are some trials using very specific cell types for specific conditions—such as embryonic stem cell–derived neuronal cells for spinal cord injury, retinal progenitor epithelial cells for Stargardt's disease that causes progressive vision loss in adolescents, and others. Most such trials are using some form of bone marrow–derived mesenchymal stromal cells (MSCs) for a wide variety of indications. In some cases, the clinical utility has been positive; in others, it has been nonexistent. One of the problems is that we still do not know exactly how these MSCs function—whether by stimulating an immune response, sending signals to neighboring cells, or turning into the desired cell type that was injured. As a consequence, many individuals are entering clinical trials without experiencing any benefit. Often performed as an autologous cell transplant in which the patient receives his or her own cells after they have been removed and treated, no harm may be done but false expectations may arise, and clinics may offer cell therapies as solutions for conditions when there is no scientific basis for doing so. As a result, the phenomenon of what is known as stem cell tourism—for example patients, many feeling quite desperate about their state, traveling to clinics and doctors around the world in search of answers to their health problems—has become a social problem. In response, not only have TV news shows like *60 Minutes* conducted exposés on certain clinics, but organizations like the International Society for Stem Cell Research (ISSCR) have also promulgated guidelines for people to consider as they contemplate undergoing procedures or entering clinical trials.

It is tempting to be dogmatic and say such trials are either all bad or all good, but it is important to consider that there are many kinds of clinical interventions with many kinds of stem cells and derivatives, and they will all have different risks attached to them. In some cases, the public is unaware of those differences; in others, scientists do not yet even know all the risks. At the same time, however, medical progress is often made through empirical testing and innovation at the bedside. One of the dilemmas we face now and will increasingly encounter is balancing the challenge between enabling clinical innovation and advancing the state of our medical knowledge with the need for controlling risks and managing patient safety and well-being. As Patrick Taylor points out in his article, "Overseeing Innovative Therapy without Mistaking It for Research" (*Journal of Law, Medicine and Ethics,* Summer 2010), innovative therapy is never, by definition, the safest course of action, and "it is important to recognize that innovative therapies and their associated risks change over time as a result of increasing knowledge and experience . . . Initially, for a paradigm-shifting innovative therapy, knowledge and experience are low, risks may be high, and benefits will be uncertain and possibly minimal. Later, as knowledge and experience increase, risks diminish, and benefits increase." In other words, how do we manage the balance and the trade-offs while pushing the frontier on medical advances with the appropriate safety oversight in a manner that recognizes that the balance point will shift over time?

In the past several years since the first edition of this book, we have seen progress on many scientific fronts including new methods for reprogramming cells, repopulating decellularized organs with stem cells, and new techniques (e.g., spray-on skin cells). The early investments made by states and nonprofit organizations around the world have created scientific advances that are starting to attract interest and investment from the commercial sector. However, there is still a gap between the two, between the basic and the commercial. Many research centers and start-up companies lament the "valley of death" issue in funding the movement of early stage, basic science to the clinic and the market. This "translational" research typically is later stage work than is funded in academic labs but is often too early or insufficiently de-risked to attract funding from corporations or from venture capitalists. In the difficult financial environment of the last few years, investors have been looking for later stage, more fully proven investments, maybe even products, that are already in human testing. As a consequence, many promising projects at this early stage of risk enter the so-called "valley of death" between research and clinical development, where they run out of resources. One of the dilemmas facing society will be how to fund such science. To date, this gap has often been filled in the United States by private philanthropy in the form of disease foundations, but those organizations do not exist in quite the same way elsewhere around the world. Rather than putting multiple fingers in the dyke and relying on ad hoc solutions, countries should consider, at the national policy level, how to fund such work programmatically. Is this the proper role of government, of academia, of nonprofit disease foundations, of venture capitalists, or of the commercial sector? Countries may and will choose different paths, depending on their social agenda and tax policies, but it is a dilemma they all face.

In short, despite the advances of the last few years, we are still far from fulfilling the promise of stem cell science, and no matter how promising that science may be, cures for the diseases that stem cells address are not right around the corner. We need to continue to advance on the scientific front and address the attendant ethical, sociopolitical, economic, and regulatory questions simultaneously. And as stem cell research moves forward around the world, these policies will need to be coordinated—both for scientific productivity and for the ultimate benefit of patients. The complexity of the science will require the community to collaborate, the importance of the ethical and social issues will require that collaboration to occur in a properly regulated framework, the immediacy and immensity of patient needs will require that we collaborate effectively and efficiently without reinventing the wheel. Our collective need to engage the hard questions that the science brings is urgent and immediate and there is no way out of that dilemma.

Brock Reeve
Harvard Stem Cell Institute
Cambridge, Massachusetts

Prologue

INTO THE CAVE

You do ill if you praise, but worse if you censure, what you do not understand.
—Leonardo da Vinci

Exactly how Richard Dalton, royal librarian for "Mad King" George III, made his startling discovery of the anatomical drawings of Leonardo da Vinci isn't known. What *is* known is that one day, around the time of the American Revolution, while removing the contents of a chest hidden away in Kensington Palace, Dalton uncovered hundreds of exquisite drawings by the Renaissance artist. Many drawings were of the human body—of muscle and bone, lungs and hearts, legs and arms, sex organs, and even fetuses in their maternal pod. Leonardo was known as much as an anatomist as a painter and an engineer, but until these drawings came to light, most of the tangible evidence was missing.

Worth many billions of dollars, Leonardo's drawings are considered the most prized holdings of the Royal Library at Windsor Castle. Yet their significance is priceless. Perhaps more than any other artifact, the anatomical drawings mark the end of the old-world order and the beginning of the new. In the millennium between the fall of Rome and the rise of the Renaissance, the knowledge of nature existed in a largely fixed state. There was no place for curiosity in the lockdown mind-set of the Middle Ages. Then came the rediscovery of classical Greek and Roman culture and the call for reform, the rise of technology and science and printing as well as international trade, the exploration overland and overseas; the expansion of art, and perhaps above all, the heretical but irresistible hunger for human progress. The Renaissance marked the passage of a world influenced by things unseen to a world influenced by things seen and understood through careful observation. Things like the human body.

Dalton's unveiling of Leonardo's anatomical drawings liberated a vital energy stored in the artist's compositions, an energy that continues to spur the human drive to discover. Never before had the body been subjected to such powerful examination. Never before had the particulars of what we are made of and how we work been rendered in such detail: how we breathe, how we move, how we sense our world, how we nourish and repair and re-create ourselves.

It is a curious quirk of history that these magnificent works of corporeal art should find their way to Windsor Castle. If you look out the windows of the Royal Library, you see Windsor Great Park, the site of sylvan reverie for the ancient Celtic people. Here, the druids dreamed of everlasting life in a land called Tir Nan Og, where pain, disease, and decay did not exist. Tir Nan Og, the Land of Forever Young, was the Celtic expression of a universal theme in mythology, a dream that flows through all cultures and most religions: a dream of immortality through regeneration. Little did they dream, so long ago, that one day the name of their mythical land of eternal youth would reappear in a scientific and medical quest to extend life.

The ancient Celtic world of regeneration and Leonardo's legacy of scientific investigation during the Renaissance have converged in a new biology in the twenty-first century. The leading edge of this new biology, this biorenaissance, and the object of its exploration, is the most important element of all life: the cell. But not just any cell. More and more, modern science and medicine is and will be the province of the stem cell. Day by day, researchers are unveiling the mystery of the stem cell and its power for regenerating tissue that is healthy and repairing tissue that is diseased or damaged.

Stem cells are proving to be the silver bullet, the Holy Grail of medicine. They could alleviate all manner of suffering, whether it's caused by disease, injury, or genetic fate. Different stem cells possess different powers. Embryonic stem cells, for instance, have the capacity to re-create and repair any of the body's tissues and organs. Scientists often call these cells the "gold standard" of stem cell research. Adult stem cells, like those present in bone marrow, have the potential to repair some tissues and organs, but not all. Skin cells and other types of mature cells can now be genetically reprogrammed into stem cells that have many of the characteristics of embryonic stem cells. There may come a day when you go to a clinic, have some blood drawn, and have the white cells in your blood specimen reprogrammed and used to form your future stem cell tissue repair bank. If recent scientific advances are confirmed and extended, it is possible to imagine that your "personalized" banked cells will be able to repair your tissues and organs. You wouldn't have to worry about your body rejecting the cells because, after all, they are your own. One of the biggest challenges will be to make cell processing for therapies, which today can be expensive and must be approved by the U.S. Food and Drug Administration, generally affordable.

Stem cells have the potential to provide new and more effective treatments for diabetes, heart disease, genetic diseases, neurological diseases like Parkinson's and Alzheimer's, and even cancer; to repair debilitating injuries, such as spinal cord damage; to restore lost function, such as our sense of sight, hearing, smell, and touch, even limbs lost in combat. Already they have enabled blind mice to see, paralyzed rats to walk, and monkeys suffering from severe Parkinson's disease to show dramatic improvement in their symptoms. Stem cells could alter the way we look and feel, whether we wish to restore hair to our bald heads or to counteract the effects of aging on our skin, bones, and cartilage. In the eyes of enthusiasts, stem cells represent the best pathway toward the elusive fountain of youth.

Because so much of human disease is genetic in origin, and because stem cells loom larger all the time in diagnosis, treatment, and prevention of disease, stem cells will change the practice of medicine forever. The fields of stem cell biology and genomics are poised to reveal your health risks long before disease strikes and be able to take steps to minimize those risks. If disease strikes, treatment will be tailored to your unique genetic makeup and biochemistry. The code used by your stem cells to build you from the ground up will be available for tissue renovation or replacement due to disease, injury, or aging.

As the age-old dream of regeneration is being animated by stem cell research, scientists and bioengineers are re-creating components of the body's various systems from raw materials that exist within. These lab-created components—a lymph node, a nerve bundle, a heart valve—are invaluable for understanding human development as well as for refining regenerative medicine and designing biologically based drugs. If the immune system, for instance, can be re-created piecemeal outside of the body and plumbed for its secrets of defense against disease and infection, stem cells indeed will have made a momentous contribution to human knowledge. Lab-recreated organs such as skin, bladders, and windpipes are already being used in patients, with bioartificial hearts, lungs, kidneys, livers, and intestine on the horizon, as we will see.

Yet depending on who you are and perhaps where you live, stem cell research can be seen as a means for preserving life or taking it. It can be seen as generating economic competitiveness or a moral decline, sustaining scientific prestige or the rise of amoral elites, ensuring personal freedom or bondage. It is both the agent of health and a widening avenue in the bioweapons arms race. It is imperative that we, as global citizens, understand the stem cell's awesome potential for life *and* death.

Advances in stem cell research and the benefits this research promise to make available to patients collide head-on with our common understanding of life, liberty, and the pursuit of happiness, or at least profit. No other raw material has in its essence the ability to alter our fundamental definition of when and how life begins, as do stem cells. No other field of scientific inquiry involves such global diversity of cultural, religious, and ethical practices as does stem cell research. No other endeavor has private industry, research universities, and governments prepared to seize stem cell technology for their own competitive economic and political advantage.

The struggle over who gets what and how, over individual rights versus the common good, over the sacred versus the profane, is part of a larger, ongoing struggle: that between tradition and progress. Circumstances in the 1850s leading up to construction of America's transcontinental railroad illustrate the point. The "Iron Horse" gave birth to regional economies across the frontier and built diverse religious communities for new settlers. Economic success and moral righteousness were earned through hard work, ingenuity, capital, and competition. At the same time, the railroad dispossessed Native Americans, exploited immigrant labor, transected ecosystems, and fed corruption. Was the moral cost of the technological marvel worth it?

Like Leonardo, we live in a time of profound transformation. The twenty-first-century biorenaissance is as far-reaching in science, medicine, and evolution as the fifteenth-century Renaissance was in art, architecture, and culture. Leonardo's was a time of flight into the artistic and scientific unknown through observation and experimentation, flight into new worlds accessible only by long voyages over monster-laden seas, flight often forbidden by spiritual and temporal authorities. Then, as now, such flights of imagination are resisted.

It is a resistance born of the eternal dilemma of hope and fear, the same dilemma Leonardo faced as a young man when, after wandering in Tuscany on a hillside after a fierce storm, he came upon the mouth of a huge cavern. As he stood in front of it, he was seized by the question of what to do—to explore or to retreat: "I had been there for some time, when there suddenly arose in me two things, fear and desire—fear of that threatening dark cave; desire to see if there was some marvelous thing within." In all of history up to that time, fear tended to overcome curiosity about what was inside the cave, what lay beyond the darkness. The "marvelous thing within," if it existed at all, wasn't worth the risk of discovering it. New ideas and new ways of doing things made people uncomfortable. But being uncomfortable is how we make advances. That tension between where we are and where we might be is the fundamental nature of human progress. To move forward, we must ask questions and encourage freedom of thought. We must invite criticism. We must enter the cave.

Leonardo loved science because of what it enabled him to do. Of science, he said, "is born creative action," an activity he valued above all others. In exploring the human body, he explored the microcosm of the great macrocosm, the whole universe, including the heavens that Galileo later charted with his telescope. Science has looked outward as well as inward ever since, at stars and galaxies, at human, animal, and plant tissues, at microscopic life forms. When in 1978 scientists looked into a petri dish and saw the beginning of Louise Brown, the world's first test-tube baby, a vital threshold in the history of human experience was crossed. A chasm opened, separating us permanently from our past—and separating some of us from others over the wisdom of proceeding into the brave new world of the future. Not only can human beings be created in a petri dish; human genetic and degenerative diseases are already being re-created in a petri dish, thanks to stem cells. Cells from patients with diabetes, heart disease, genetic disease, neurodegenerative disease, and other diseases have been reprogrammed in the laboratory to serve as testing beds for future therapies. That means treatments will be based on knowledge derived from the patients' own cells and genetic makeup. They are the best barometers for effective treatment.

Those of us who hope to benefit from the bounty that stem cells have to offer also play a role in shaping the future. First, we must learn what the experts can teach us about the power of stem cells. We need to know what stem cells are and how they work. We need to know when and whether it's right to use them and how to regulate that use—ethically, politically, and competitively. We need to understand that stem cells are agents of hope but also harbingers of destruction. Our learning curve is steep and the language isn't always clear. But the more we know about stem cells, the more we realize how high the stakes are in deciding their fate.

It is we the people who must make this decision—in the court of public opinion and in the sanctity of the voting booth—so that it is not made for us.

Sigmund Freud once described Leonardo da Vinci as a man who awoke too early in the darkness while everyone else was still sleeping. British art historian Sir Kenneth Clark called him "the most relentlessly curious man in history." Who better than the original Renaissance man to inspire us, to rouse us to see the brilliant possibilities of the biorenaissance and its leading light, the stem cell? In the right hands, stem cells shine like a beacon of hope. In the wrong hands, they threaten to extinguish humankind for all time. Their fate is our fate. We must awake early, to all of the possibilities of the stem cell, and remain awake.

Our future depends on it.

Chapter One

AGENTS OF HOPE

Knowing is not enough; we must apply. Being willing is not enough; we must do.
—Leonardo da Vinci

Toward the end of his life, Leonardo da Vinci was a guest of King Francis I of France. Leonardo lived in the manor house at Cloux near the king's Amboise chateau in the Loire Valley. It was a pastoral setting that gave the artist time to reflect, put his notebooks and drawings in order, and make out his last will and testament. It is recorded that he asked his assistant Francesco Melzi to get a treatise for him, *On the Formation of the Human Body in the Mother's Womb*, by the thirteenth-century theologian Giles of Rome. Now, at the end of his life, Leonardo was thinking about how life begins.

Human reproduction had always fascinated him. In 1512, at the height of his powers, he drew *The Fetus in the Womb*. "The womb, split open like a burst seed-case, reveals the coiled fetus, shaped into compelling roundness by the rhythmic curves of his pen," wrote Leonardo authority Martin Kemp in *Nature*. One of his most famous anatomical drawings, it depicted what he called "the great mystery," a mystery in many ways more profound than the enigmatic smile on his *Mona Lisa*. With *The Fetus in the Womb,* the study of science, medicine, and human reproduction were brought to bear on that mystery. "The navel is the gate from which our body is formed by means of the umbilical vein," he wrote. What Leonardo could not have imagined as he examined the umbilical cord attaching the fetus to the mother was that it is a treasure trove of stem cells—cells with regenerative powers that someday may eradicate any number of diseases. Stem cells that already have saved the lives of people with diseases of the blood and bone marrow. People like Molly Nash.

Like most girls her age, Molly Nash loved to dance. But that was before 2000, when the Colorado child began suffering from the effects of Fanconi anemia, a genetic disease that causes catastrophic failure of the bone marrow. Bone marrow makes life-saving blood cells: white blood cells that fight infection, red blood cells that carry oxygen to organs and tissues, and platelets that produce blood clots to stop bleeding. As their bone marrow deteriorates, people with Fanconi anemia suffer extreme fatigue, frequent infections, nosebleeds, and bruises. Those who survive into adulthood are at risk for developing a host of cancers. By the time she was six, Molly had already been diagnosed with an early form of leukemia.

Molly's parents, Lisa and Jack Nash, set out to save their daughter with the help of John Wagner, a pediatrician at University of Minnesota Medical Center in Minneapolis, director of the Division of Hematology-Oncology and Blood and Marrow Transplantation, and codirector of the Center for Translational Medicine. Wagner proposed a rescue plan never before attempted. On the surface, his plan sounded simple: doctors would replace Molly's diseased blood with healthy blood. Problem was, the healthy blood needed to be a perfect match. Furthermore, the perfect match had to be Molly's sibling, a newborn whose umbilical cord blood contained the stem cells that could create the healthy blood. Since Molly had no such sibling, her parents had to conceive one. In fact, to make sure the new sibling's blood would be free of Fanconi anemia, the Nashes had to create a dozen embryos through in vitro fertilization from which the perfect match could be chosen via genetic testing. Finally, nine months after Lisa was implanted with the chosen embryo and delivered a healthy infant, Molly would receive a bone marrow transplant using the blood-forming stem cells from her new sibling's umbilical cord.

The plan worked. As Molly held her newborn brother, Adam —dubbed "the world's first designer baby" by a national news program—his donated stem cells made their way to her bone marrow and set about rebuilding her entire blood system with healthy cells. Three weeks later, she started to dance again. Eleven years later, she is in good health.

Molly's rescue marked the first time that PGD—preimplantation genetic diagnosis—was used specifically to ensure a perfect donor of umbilical cord blood stem cells for transplantation. PGD offers hope not only to patients with Fanconi anemia, but those with leukemia, thalassemia, and other blood diseases that cause the immune system and bone marrow to fail. Because PGD can determine whether an embryo is male or female, the technique can also reveal sex-specific blood diseases like hemophilia. Said Molly's physician, John Wagner, now one of the nation's leading authorities on umbilical cord blood transplantation, "Molly is an example of how the work done to combine preimplantation genetic diagnosis and in vitro fertilization to create a healthy cord blood donor holds great promise."

The hope of medicine based on the regenerative powers of the stem cell is a powerful hope. Perhaps not since the time of

Hippocrates has there been reason for scientists, physicians, patients, and their families to have such hope. No new approach to dealing with the monumental suffering and social costs of major diseases comes close to the promise of stem cell therapy. That promise includes children who suffer from diabetes or genetic disorders like Hurler's syndrome, cystic fibrosis, Tay-Sachs disease, Batten disease, Marfan syndrome, and muscular dystrophy. The promise includes middle-aged adults who suffer from heart disease, Lou Gehrig's disease, multiple myeloma, or spinal cord injury. The promise includes aging adults who suffer from Alzheimer's and Parkinson's disease and macular degeneration, which can lead to blindness. The stem cell could even turn the current understanding of how cancer begins, and how to treat it, on its head. It is the power of hope in regenerative medicine that propels some patients to seek unproven treatment around the globe—in China, India, Thailand, Russia, Ukraine, Mexico, Costa Rica, the Dominican Republic, Germany, Portugal, the Netherlands, Argentina, and Brazil. In some countries, notably China, such "stem cell tourism" has become a multimillion dollar industry.

Though most people probably don't realize it, the promise of stem cells as a successful therapy in modern medicine is nearly forty years old. It came in the guise of bone marrow, where blood-forming stem cells, like Adam Nash's, set up shop early in embryonic development.

DISEASES AND CANCERS OF THE BLOOD

The first successful bone marrow transplant in humans took place in 1968 at the University of Minnesota Hospital in Minneapolis. The patient was a four-month-old boy suffering from a life-threatening immune deficiency that had already claimed his brother. The donor was the patient's sister, whose bone marrow supplied the blood-forming stem cells that replaced the boy's diseased cells and restored his immune system to health. For the first time, a human body had accepted bone marrow from someone other than an identical twin—someone whose tissue was nevertheless a good match for his own. That boy is now a healthy forty-three-year-old father of twins.

Leading the transplant team was Robert Good, one of the first scientists to view the immune system as a coordinated, complex system rather than a collection of piecemeal blood and tissue components. A year after Good's feat, E. Donnall Thomas performed the first successful bone marrow transplant to cure leukemia. Thomas became known to many as the father of the bone marrow transplant, building the Fred Hutchinson Cancer Center in Seattle into the world's largest bone marrow transplant program. For his innovations in science and medicine, he was awarded the 1990 Nobel Prize.

Before the pioneering work of Good and Thomas, diseases and cancers of the blood that destroy the immune system were a virtual death sentence. Since then, those who suffer from anemia, leukemia, lymphoma, and multiple myeloma have hopes for survival because of bone marrow transplants and blood stem cell transplants. Every year, about 50,000 such transplants are done worldwide. Thirty thousand are done using the patient's own blood-forming stem cells, a procedure called autologous transplants. Another 16,000 procedures are done using bone marrow from a donor who is unrelated but genetically matched as closely as possible; these are called allogeneic transplants. Of the more than 7,000 allogeneic transplants performed in North America in 2002, more than 5,000 were for leukemia or preleukemia. More than 9,000 of the 10,500 autologous transplants performed in the same year were for multiple myeloma, an aggressive cancer of plasma cells that make blood antibodies, or lymphoma, a cancer of the lymph system.

Multiple myeloma kills 11,000 Americans each year. It killed columnist Ann Landers and actor José Ferrer. Former U.S. Congresswoman Geraldine Ferraro, the first woman to be a candidate for U.S. vice president of a major American political party, was diagnosed with multiple myeloma in 1999 and survived twelve years. In late 2003, cancer researchers discovered that a renegade stem cell trafficking among the plasma cells of the bone marrow starts a killer clone that gives rise to multiple myeloma. The renegade cell can survive aggressive chemotherapy and mount a comeback. This means a relapse for far too many persons with the disease. Treatments for multiple myeloma, including thalidomide and other drugs, have not improved survival rates very much in the last twenty years, but transplantation of stem cells recruited from the patient's circulating blood is showing real promise.

The story of the transplantation of stem cells collected from circulating blood as a valuable clinical therapy begins with Irving Weissman, director of Stanford University's Institute for Stem Cell Biology and Regenerative Medicine. It was Weissman who first isolated specific blood-forming stem cells in mice, in 1988. Just four years later, Weissman and others isolated the human counterparts of the mouse stem cells. The story picks up with James Thomson of the University of Wisconsin in Madison, the scientist who made the first line, or family, of human embryonic stem cells in 1998. Thomson reported on September 11, 2001—coincidentally the day of the terrorist attacks on New York and Washington, D.C., and the appeal for blood donation that followed—that his team could "direct" embryonic stem cells to form colonies of the normal cells found in blood. The development of stem cell–based blood could have implications for human medicine far beyond the treatment of blood malignancies. Stem cell–based blood also could be used to thwart rejection of transplanted foreign tissue in treating diabetes, Parkinson's disease, or spinal cord injury as well as in traditional organ transplantation. Stem cell–based blood could create a limitless supply of donor blood, eliminating the need for donations in times of massive emergency requests, such as 9/11. Stem cell–based blood could even replace blood destroyed by a radiation-generating nuclear weapon. Radiation kills rapidly dividing cells, including blood-forming stem cells. Big challenges remain making this a reality, however.

It is thought by some medical scientists that umbilical cord blood harbors adult stem cells that have many of the important attributes of embryonic stem cells. In other words, adult stem cells from cord blood can form diverse tissue cell types *other* than those that form blood. Someday, when the need arises for tissue regeneration—whether it's cartilage for the knee, muscle for the heart, or neurons for the brain—perhaps scientists will have figured out how to reprogram these particular adult stem cells to generate the necessary tissue to repair the problem. Much more research will need to be done before that day arrives.

In the meantime, more and more parents are storing the umbilical cord blood of their offspring in the event of future need—in case disease should strike that is treatable with blood stem cells. More than thirty private cord blood banks were in operation in 2010 in the United States, up from a mere dozen in 2000. Companies like StemCyte in the United States and Smart Cells in the United Kingdom process the blood-forming stem cells, then freeze them in liquid nitrogen at a cost of up to $2,000 per child. Cord blood stem cell registries are also expanding in the not-for-profit sector with several initiatives, including an $8 million investment by the National Marrow Donor Program in 2004. The Germany-based International NetCord Foundation is a network of nonprofit public cord blood banks in the United States, Europe, Israel, Japan, and Australia. These developments make available potential lifesaving blood-forming stem cells at their basic cost with nominal profit. In 2006, the U.S. Health Resources and Services Administration began awarding contracts to public cord blood banks to increase the supply of donations. "We can find donors for everyone," the University of Minnesota's John Wagner told the Associated Press.

HEART DISEASE

Five hundred years after Leonardo da Vinci's intricate drawings of the heart and heart valves, drawings that continue to inspire heart surgeons today, cardiovascular disease is a massive and growing worldwide public health problem. It is estimated that it kills more than 800,000 people in the United States each year, making it the leading cause of death, with another 140,000 dying of stroke each year. Congestive heart failure—the ineffective pumping of the heart caused by the loss or dysfunction of heart muscle cells— afflicts nearly five million Americans, with 400,000 new cases each year. In 2005, more than seventeen million people died from heart disease and stroke around the world. That number represents 30 percent of all deaths from disease and injury, according to the World Health Organization. And no longer is heart disease primarily a problem of affluent countries with their characteristic high-fat diets, bulging waistlines, and red-zone levels of stress. In fact, heart disease is growing at a rate faster in certain parts of the developing world than in the first world. In India, the country with the world's second largest population, heart disease is a huge health problem. Many Indians think that heart disease is "a 'Rich White Man's disease,'" Ivan Berkowitz, a scholar at the International Academy of Cardiovascular Sciences, told the *Times of India* during a conference in 2011. The academy calculates that 80 percent of deaths due to heart disease—thirty-five people every minute—occur in emerging nations.

The "Race Against Time" study by Columbia University's Earth Institute estimated that more than twenty million years of future productive life are lost annually due to cardiovascular disease. The study showed that middle-aged men and women in Brazil experience mortality rates from heart disease that were similar to those in the United States thirty years ago. Today, Brazilian death rates are 40 percent higher for men and 75 percent higher in women with heart disease within the same age group compared to the U.S. rates. If current projections hold, and preventive disease management isn't implemented, Brazil will experience a 15 percent increase in its heart disease–related death rate per decade for the next thirty years.

While Brazil works to develop a comprehensive strategy to prevent cardiovascular disease, a decade ago, its scientists and clinicians started pioneering the use of adult stem cells to treat heart failure, a common consequence of chronic heart disease. Scientists in animal laboratories had previously demonstrated that blood-forming stem cells could generate new capillaries to deliver blood to tissue damaged by a heart attack. In 2003, Hans Dohmann at Hospital Pró-Cardíaco and Federal University of Rio de Janeiro, in collaboration with Emerson Perin and his colleagues at Houston's Texas Heart Institute, used a mixture of adult stem cells from bone marrow to treat fourteen Brazilians with severe heart disease. Each patient's damaged heart muscle was injected with thirty million stem cells drawn from their own bone marrow. Two months later, the treated patients had significantly fewer symptoms of heart failure and a greater ability to pump blood than the untreated patients. After another two months, the treated patients showed even more improvement, with stable cardiac pumping activity and no irregular heart rhythms. "Either these stem cells became new blood vessel and new heart muscle cells, or their presence stimulated the development of one or both" from something within the ailing patients' hearts, wrote James Willerson, president of the University of Texas Health Science Center at Houston and chief of cardiology and medical director at the Texas Heart Institute, in summarizing the study's results. Willerson later added a caveat: "Stem cells here have to be considered in quotations because we've taken bone marrow cells and separated them into cells with a single nucleus, some of which are stem cells and some of which are not." Nevertheless, the results were persuasive enough that in early 2004, the U.S. Food and Drug Administration approved one of the first clinical trials in the United States to test a bone marrow stem cell therapy for severe heart failure.

Even as Perin and his colleagues at the Texas Heart Institute began planning the new study, results from the first randomized trial of adult stem cell injections in heart failure patients were being reported at the annual meeting of the American Association for Thoracic Surgery in 2004. Cardiac surgeon Amit Patel from the University of Pittsburgh School of Medicine and the McGowan Institute for Regenerative Medicine, and colleagues from the United States and the Benetti Foundation in Rosario, Argentina, reported that transplantation of bone marrow–derived blood stem cells can be a viable treatment for congestive heart failure because they promote growth of blood vessels and heart muscle. Ten Argentine patients received injections of bone marrow–derived stem cells directly into the damaged muscle of their heart during heart bypass surgery; ten others received only the bypass operation. Six months later, the hearts of the group treated with stem cells were pumping more blood than those who had the bypass operation only. Their approach could revolutionize treatment of heart failure from one of relieving symptoms to one that is "truly regenerative," said Patel's Pittsburgh colleague, Robert Kormos.

The explosion of interest in stem cells across the globe has detonators in many fields of medicine, but in none are the possibilities more profound or the need for success greater than in cardiology. Yet funding for clinical trials using stem cells in heart disease is extremely low. Compared to trials being done to test traditional and genetically engineered drugs, the industry has little experience with cellular-based therapies. There is also the matter of money down the line. If a patient's own cells may be his or her best therapy, costs could be limited to cell processing, the procedure itself, and the hospitalization and perhaps not require expensive drugs or biopharmaceuticals. This scenario of manageable costs for processing a patient's own cells, however, seems unlikely. A "universal donor" stem cell might be developed, or a stem cell bank might be established, to provide treatments for large segments of the population affected by heart disease. The cells a patient needs would be "matched" to appropriate cells donated to the bank, much like what is done today in bone marrow, kidney, and liver transplantation.

Whether adult stem cells will indeed, in the end, prove to be both safe and effective for treating heart failure won't be known anytime soon. So far, human and animal research programs in South America, Europe, South Africa, Asia, and the United States have yielded inconsistent, even conflicting results. Some patients show a slight increase in cardiac output when treated with their own bone marrow or heart muscle stem cells. Only recently have serious attempts been undertaken, however, to standardize the specific cell type used in treating patients. Indeed, the variety of cells derived from bone marrow used in some of the early studies was described as a witch's brew of bone marrow cells. A large national stem cell study for heart disease showed that transplanting a purified type of bone marrow stem cells into the heart muscle of subjects with severe angina—chest pain or discomfort that occurs when an area of your heart muscle doesn't get enough oxygen-rich blood—resulted in treated subjects experiencing less pain than the control group. Such studies need to be confirmed and extended.

Bone marrow stem cells were first used to treat a heart patient by Bodo-Eckehard Strauer and his team at the University of Düsseldorf in Germany. After success with his first patient, Strauer expanded his study, which was published in 2002, and other German teams chimed in with their own studies. That dismayed some U.S. scientists who saw the German studies as premature. Typically, such procedures are tested in large mammals after testing is done in mice and rats. Within months, more than a dozen trials in humans were under way in France, Brazil, and South Korea, but not in the United States because of the FDA's more stringent demands for safety. Surely among the FDA's concerns is the fact that benign tumors can form when embryonic stem cells are injected into experimental animals. There is also a risk, albeit a very small one, that these benign tumors could transform into a malignant cancer. Would the benign tumors form in humans and then transform to malignant ones?

Well-intentioned and carefully planned studies on patients may go awry. Some of the setbacks might be severe, but waiting to answer the questions posed by biology has its costs, to individuals and societies. While we have been very successful in treating cardiovascular disease, the growth of heart drugs sales, cardiac surgery, and biomedical engineering cannot compete, on a global scale, with the growth of what they are designed to treat. These advances are running up against "the emerging demographic profile of the world, where every nation is now facing a change in the age structures of its population," according to Earth Institute director Jeffrey Sachs.

If stem cell regeneration of damaged heart tissue does become a clinical reality, it could transform the treatment of one of the world's leading health care problems, congestive heart failure. By late 2006, more than four hundred heart patients worldwide had been treated in human trials using the marrow cell-to-heart approach, with only modest success in the view of most cardiologists. Could this be the hope of the future for the millions of people worldwide with cardiovascular disease or with end-stage heart failure, a rare few of whom wind up tied to artificial hearts? While there have been no reports of serious adverse effects, the beneficial effects have been modest. The research community is somewhat divided about the value of the approach, ranging from skeptics like Stanford's Irving Weissman to enthusiasts like Douglas Losordo, whose research team at Northwestern University reported following a large clinical trial that heart patients treated with their own purified bone marrow stem cells experienced less severe pain associated with angina and improved exercise capability. In his short introduction to stem cell research, developmental biologist Jonathan Slack put it this way: "Scientists generally regard the whole enterprise as pointless, lacking rationale and producing no valid results. Clinicians cite the statistically significant positive results and justify the activity on the grounds of benefits to patients, while admitting that the mechanism is not understood."

So there is as yet no definitive answer to the question of whether adult stem cells can help some of the millions of people with advanced heart disease. As we noted above, most studies to date using bone marrow stem cells have used mixtures of cells, and until pure populations of stem cells are used in large human clinical trials, we really won't know. As Britain launched an $80 million "mend broken hearts" program in 2010, the *Financial Times* reported that an alternative to cellular therapy picked up support. Paul Riley, a professor of molecular cardiology at University College London, said the ideal would be to find a cocktail of small drug molecules that mobilize a patient's own heart stem cells to repair and regenerate damaged tissue.

Broken hearts wouldn't need to be mended if hearts could be regrown. That prospect is not science fiction. In late 2006, Swiss scientists used stem cells from amniotic fluid to grow human heart valves in the laboratory, valves that functioned normally when tested. University of Minnesota biomedical scientist Doris Taylor and her colleagues seeded the protein framework or "skeleton" of a rat heart with stem cells and progenitor cells. The cells regenerated heart muscle and vessels, and the heart started to beat with the aid of a pacemaker. Their report, published in *Nature Medicine* in early 2008 and entitled "Perfusion-decellularized matrix: using nature's platform to engineer a bioartificial heart," was greeted with great fanfare by the news media. Taylor was widely quoted as saying, "Give nature the tools and get out of the way." The "tools" are cells including stem cells that have an ability to find their proper location in the geography of an organ, divide and multiply to fill in the space, and begin to carry out the task nature assigned to them.

Taylor's team "decellularized" or removed cells from the rat heart by dripping a mild detergent through its vessels. Over time the detergent broke apart all the cells in the organ and washed away the cellular material, leaving behind a "ghost heart" consisting of a protein framework and a treelike vascular system wrapping around and branching within the framework. They then put the heart into a bioreactor—a chamber that supports a biologically active environment—and seeded the white translucent heart framework with a variety of types of heart cells from newborn rats. The cells found their way to their customary location—heart muscle cells to the heart walls and chamber, endothelial cells to the heart's vessels—where they began to divide and fill in the framework. After eight days, the heart responded to electrical stimulation by contracting rhythmically and meeting numerous benchmarks for typical cardiac function. "Though the current study is limited to rat hearts," the authors wrote, "this approach holds promise for virtually any solid organ." Since Taylor's team reported its breakthrough, other research teams at Harvard, Yale, Wake Forest, Minnesota, and other universities used the perfusion decellularization method to re-create functioning animal livers and lungs. Researchers at the University of Florida used the technique in conjunction with mouse embryonic stem cells to re-create a functioning mouse kidney. Kidneys are the most commonly transplanted organ. More than 16,500 kidney transplants were performed in the United States in 2008. Meanwhile, Taylor and her team are seeking to have the body's master cells restore the architecture of "ghost hearts" from human cadavers and pigs and then induce the rebuilt organs to beat.

Re-creating complex human organs through decellularization and stem cell recellularization and then transplanting retrofitted organs into patients is still years away. First it has to be done successfully in laboratory animals, as it has, but retrofitted organs have been tested in animals only for a few hours at a time. Taylor concedes the challenge for biologists, bioengineers, and physicians. "We are a long way off creating a heart for transplant," she told the London's *Sunday Times*, "but we think we've opened a door to building any organ for human transplant."

DIABETES AND OTHER AUTOIMMUNE DISEASES

The most powerful symbol of hope on the landscape of regenerative medicine may be the five-ton granite monument in front of the Frederick Banting Museum in London, Ontario. The monument houses the Flame of Hope, kindled in 1989 by Queen Elizabeth, the Queen Mother. The flame will be extinguished when a cure for diabetes is found.

Until then, January 11, 1922, remains the "day of hope" for the millions of people worldwide with diabetes. On that date, Frederick Banting and Charles Best at the University of Toronto began using an extract of pancreas to treat Leonard Thompson, a deathly ill fourteen-year-old diabetic who weighed only sixty-five pounds. The results were disappointing, but six weeks later they tried a more refined extract from a colleague, biochemist James Collip. Soon after the youngster received "Collip's extract," his blood sugar returned to normal. The extract was called insulin, and the discovery was one of the greatest medical achievements of the twentieth century. It was the first effective treatment for diabetes, allowing those with the disease to view it more like a chronic condition than a death sentence.

More people are living with diabetes today because of insulin than at any time since Egyptian physician Hesy-Ra first described the disease in 1552 BC. But insulin is not a cure for diabetes. People with the disease, particularly those in advanced industrial societies where it's most prevalent, still face a reduced quality of life and the increased likelihood of complications such as heart disease, stroke, kidney failure, blindness, and limb amputation. A cure is a matter of great importance, even to those not directly affected by diabetes. The price tag for treating the disease in the United States was $116 billion in direct costs in 2007, with another $58 billion in indirect costs from days of work lost and permanent disability. If you live in the United States, your health insurance and taxes (some of which go to Medicare) are higher than they would need to be if diabetes were cured or the cost of treating it were significantly reduced.

But the truly staggering costs for treating this most common of metabolic disorders are yet to come. According to the World Health Organization, an estimated 30 million people worldwide had diabetes in 1985. An estimated 171 million had it in 2004, and at least 366 million are projected to have diabetes by 2030, with most of the increase occurring in developing countries. The 2004 World Health Report showed that diabetes kills a million people per year, nearly 2 percent of the estimated annual 57 million deaths from disease or injury worldwide. That percentage of deaths due to diabetes is expected to grow significantly.

As obesity, and especially childhood obesity, becomes more common, what is called type 2 diabetes and its complications will soon become a health care crisis for this and other countries. "We have truly a global epidemic which appears to be affecting most countries in the world," said Philip James, chair of the International Obesity Task Force, following a study showing that the number of overweight children worldwide will increase significantly by the end of the decade. Evolutionary biologist and physiologist Jared Diamond suggests that genes and diabetes epidemics may help explain the disparities in diabetes among ethnic groups. As living standards and food supplies improve in the rest of the world today, ethnic populations that still carry genetic susceptibility to diabetes will become vulnerable, he wrote in *Nature*. For some people, type 2 diabetes is the downside of a growing food supply and affluence.

With numbers like these, it is no wonder that diabetes is almost always mentioned as a disease for which stem cell research holds the greatest promise. Diabetes research itself has been an enormous global endeavor for decades. Patient advocacy groups, particularly the Juvenile Diabetes Research Foundation, are among the largest and most active, carrying clout that no legislator can afford to ignore. Stem cell research stars like Harvard's Douglas Melton have children with the disease or know people who do. The same holds for Hollywood stars like Dustin Hoffman, Harrison Ford, and Warren Beatty, and moviemakers like Jerry and Janet Zucker. It holds for Robert Klein, the wealthy Palo Alto developer who spearheaded California's $3 billion ballot initiative in 2004 to make the state an oasis for human embryonic stem cell research.

The major impetus for efforts to enlist stem cells in the cause of a cure for diabetes is fueled by type 1 diabetes, or what used to be called juvenile-onset diabetes, the kind that typically strikes children and young adults. Type 1 diabetes has grown dramatically since 1950, with nearly four children per one thousand in Western countries requiring insulin treatment by the age of twenty.

It is the fate of some diabetic children to die in their sleep. The parents of twelve-year-old Emma Klatman, the 2003–2004 American Diabetes Association's national youth advocate, know that only too well. "Eight times a day and sometimes more, and once or maybe twice each and every night, we poke our daughter's soft fingers to find out if she needs insulin or a cookie for 'lows,'" they wrote in a letter in the *Los Angeles Times*. People with Type 1 diabetes need daily injections of insulin to live. That's because it is an autoimmune disease, in which the body's immune system attacks the tissues it's supposed to protect. In type 1 diabetes the body attacks the cells in the pancreas that make insulin. Other autoimmune diseases include multiple sclerosis, some forms of thyroid disease, lupus, and rheumatoid arthritis.

The quest for treatment of type 1 diabetes, and the monumental endeavor to counter its devastating consequences, keeps expanding. Pancreatic islet cell transplantation from cadavers via the Edmonton Protocol, a procedure developed in Canada that can restore normal blood sugar in diabetics, has been the most promising avenue for treatment in recent years. But since pancreas donations are relatively few, pancreas transplants or islet cell transplants will never be an effective strategy for eradicating or even seriously mitigating the disease. Another possible strategy, being used to treat other autoimmune diseases as well, is to re-create a patient's immune system using their blood-forming stem cells. In a small study reported in early 2007, twelve of fifteen Brazilian patients with type 1 diabetes were able to stop taking insulin soon after transplantation of their own stem cells and the rebuilding of their immune systems.

The best prospect for a cure is a renewable source of insulin-producing cells. The best hope of finding that cure is stem cell research. All major types of stem cells are under scrutiny in the effort to find a cure for diabetes: adult stem cells from bone marrow and umbilical cord blood, adult cells that have been genetically reprogrammed to behave like embryonic stem cells, and embryonic stem cells derived from the early human embryo. Both adult and embryonic stem cells are being studied with the goal of programming or reprogramming them to create the cells that make insulin. Compared to cells from cadavers, embryonic stem cells "are less likely to have accumulated cellular and chromosomal aberrations," wrote Ken Zaret of the Fox Chase Cancer Center. How embryonic stem cells might be turned into insulin-making cells is being investigated by Joseph Itskovitz-Eldor of the Technion-Israel Institute of Technology in Haifa. He and his colleagues coaxed human embryonic stem cells into forming islet-like cells that make insulin. In late 2006, the California biotech firm Novocell, later renamed ViaCyte, announced similar results and commenced preclinical testing, an essential step to complete before human trials can begin. Not to be outdone, another California company, Geron Corporation, reported its own breakthrough in the journal *Stem Cells* in 2007. Working with Canadian scientists who specialize in the Edmonton Protocol mentioned above, Geron scientists showed that they could produce isletlike cell clusters from human embryonic stem cells. The cell clusters produced insulin in lab dishes.

In 2008, Geron showed that the cells had successfully engrafted in diabetic mice where they expressed key pancreatic islet proteins and responded to high levels of glucose in the blood by producing insulin. The same year Douglas Melton and his team at the Harvard Stem Cell Institute reported the successful conversion of pancreatic cells directly into cells that closely resembled insulin-producing beta cells, showing the versatility of the reprogramming technique pioneered by Japanese stem cell scientist Shinya Yamanaka. Indeed, direct reprogramming of adult cells to a desirable cell type—heart muscle cells, neurons, retinal cells, beta cells—in a single stroke without taking the adult cells back all the way to pluripotency has caught on with stem cell researchers.

Melton and his team followed that research with another successful experiment using skin cells from type-1 diabetes patients: They reprogrammed the cells to a pluripotent state, directed them to differentiate into beta cells, and then tested whether the reprogrammed cells could function like normal insulin-producing beta cells by exposing them to glucose in a dish. The test showed that high sugar levels induced the cells to produce more of a protein that beta cells release when they metabolize sugar. At low sugar levels, cellular production of the key protein dropped off. "These cells represent the newest model of diabetes for humans," Melton told *Time*. "We have a lot of good models of type 1 diabetes in the mouse, but everything that we have learned from them has failed in the clinic. Now we have a chance at figuring out how humans get the disease." A similar model for type 2 diabetes was developed by the San Francisco biotech company iPierian, which is developing an assay to screen drugs that can be used to treat the disease.

George Daley of Boston's Children's Hospital and Harvard Medical School observed that science is often advanced by a bold new technology or critical biological reagent that makes cells do what they hadn't done before. Programming embryonic stem cells or reprogramming adult cells to form clusters of insulin-producing betalike cells makes them do what they hadn't done before. The next step—the next technical barrier to be overcome or missing link to be found—is to induce pluripotent stem cells or beta precursor cells to form fully mature and fully functional beta cells, which make lots of insulin in the petri dish and when they are implanted in the body. When that happens, the "Flame of Hope" may be down to its final days.

SPINAL CORD INJURY AND NERVOUS SYSTEM DISEASES

Until his death in October 2004, the most publicly visible person with spinal cord injury was also arguably the world's leading stem cell research advocate: actor Christopher Reeve, who was paralyzed in 1995 after being thrown from a horse. The most visible person with ALS, a type of motor neuron disease also known as Lou Gehrig's disease, is theoretical physicist Stephen Hawking. Hawking, author of the international bestseller *A Brief History of Time,* who holds a prestigious chair in mathematics at Cambridge University once held by Sir Isaac Newton, uses sophisticated electronic technology in his daily life to mitigate the consequences of his paralysis, as did Reeve. And like Reeve, Hawking is a strong advocate of biological research, in particular research that will bring about brain stem cell implants and other treatments for motor neuron diseases like ALS, spinal cord injuries, and neurodegenerative diseases like Parkinson's and Alzheimer's, all of which involve the death of nerve cells and disruption of neural networks.

It wasn't so long ago that the loss of nerve cells in the central nervous system of vertebrates was considered irreversible. That was still the case in 1939, the year that "the luckiest man on the face of the earth" gave his farewell speech to fans at Yankee Stadium. The man was Lou Gehrig, and he had just been diagnosed with amyotrophic lateral sclerosis. Two years later, the famed baseball player was dead, his name forever linked with the motor neuron disease that robs muscles of all normal function, from running and swinging a bat to breathing and swallowing. About 30,000 Americans have ALS, with 5,600 newly diagnosed each year. It typically strikes people between the ages of forty and seventy and is usually fatal three to five years after diagnosis, although advances in treatment mean that many people with the disease can live longer, Stephen Hawking included.

Today, the production of new nerve cells, or neurogenesis, is recognized as an ongoing activity at a low level in the brains of adult vertebrates, including humans. The potential of stem cells to regenerate nerve cells is being investigated around the world. Up the English coast from Hawking's offices at Cambridge University, in the Scottish heartland of the Celtic myth of Tir Nan Og, the research team that shocked the world in 1997 by cloning Dolly the sheep has set out to tackle the biology and pathology of ALS and other motor neuron diseases with cellular reprogramming technologies. Meanwhile, scientists like Fred Gage at the Salk Institute in La Jolla, California, who was the first to isolate a multipotent, self-renewing stem cell from neurons, are trying to understand exactly how stem cells give rise to neural cells and to cells that produce blood vessels to supply them. Other neuroscientists are trying to regenerate nerve cells for eventual use in the clinic, including Sally Temple of Albany Medical College in New York. Temple and her associates were among the first to find stem cells in embryonic brains of mice and to show that these cells could be grown in abundance, at least in the laboratory. Their technique was widely viewed as a major step on the path to developing treatments for people whose nervous systems are damaged or diseased.

The potential of a nerve cell to be replaced raises several questions. If with aging we are losing neurons and a mechanism exists to increase their number, does the process of cell death with aging occur at a faster rate than the ability of cells to regenerate? Do the rates of cell death and degeneration vary in individuals? If the process exists, how efficient is it? Does it carry on with aging? Does it stop in some people? If so, why? And when?

Recent evidence is provocative. It suggests that bone marrow cells and primitive muscle cells can "morph" into cells that have most and perhaps all of the characteristics of nerve cells. Su-Chun Zhang and his research team at the University of Wisconsin–Madison reported in early 2005 that they had created functional motor nerves or neurons from human embryonic stem cells. It was a complicated process, with numerous growth factors needed to coax the cells at specific times in their development. Yet in time the cells might serve as "replacement motoneurons for applications in patients with motoneuron diseases or spinal cord injury," said Zhang. With the advent of cellular reprogramming of adult cells to pluripotency in 2007, the prospect of creating an ALS disease model in the laboratory was within reach, and it didn't take long for researchers to accomplish the feat. In 2008, Kevin Eggan and his colleagues at the Harvard Stem Cell Institute described how they reprogrammed skin cells from an elderly ALS patient into a pluripotent state and then successfully directed the cells to differentiate into motor neurons, the type of cell destroyed in ALS. In theory, these motor neurons could be transplanted back into the patient to restore lost function.

But that's far in the future. For now, nerve regeneration studies in experimental animals have produced some encouraging results, prompting outspoken advocates like the late Christopher Reeve to put the task of translating promising laboratory results into human trials on the fast track. Scientists know that injecting paralyzed rats with human stem cells can restore some movement and function. But researchers want to know exactly what mechanisms are responsible for the improvements. One such researcher is Douglas Kerr of the Johns Hopkins School of Medicine. He asked why rats, paralyzed by a poliolike virus, regained some ability to move their feet after being transplanted with rat embryonic stem cells that had been coaxed into becoming motor neurons. A series of experiments later gave Kerr the idea that the stem cells had migrated to the sites of the most severe damage along the rats' spinal cords and secreted agents that rescue dying neurons and re-form enough connections so that the animals were able to move a little. In June 2006, major news media filled the airwaves with the story of how Kerr had used embryonic stem cells to engineer fully functioning neuron circuits in paralyzed rats. Television viewers saw paralyzed rats moving their limbs.

Neuroscientist Leonardo da Vinci, who pioneered systematic studies of the brain and nervous system, drew the spinal cord as a tube for the "passage of animal powers." For centuries afterward, physicians and scientists looked at severed spinal cords as a sort of medical black hole. No matter how much imagination and energy were invested in its study, nothing would come of it. Among the growing number who beg to differ today is UCLA neuroscientist V. Reggie Edgerton who, among other things, identified the importance of the cerebellum in swinging a baseball bat well, one of the most demanding tasks in all of sports. In that study, Edgerton followed in the footsteps of Columbia University psychologists who talked New York Yankee legend Babe Ruth, Lou Gehrig's teammate, into undergoing a battery of sensory and cognitive tests after a day's work in the batter's box. Physician and best-selling author Jerome Groopman described in *The New Yorker* Edgerton's determination to move the "graveyard of neurobiology" forward. Edgerton's revolutionary idea, championed by Christopher Reeve, is that the spinal cord can function more or less independently of the brain. Edgerton has spent the last two decades trying to show that memory circuits in persons with spinal cord injury can be reactivated through rehabilitation. That vision formed the basis for Reeve's extensive and unrelenting rehab program, a program that yielded a sequence of functional motor improvements that were tracked by MRI brain imaging. It should be noted that brain imaging itself is a direct descendant of Leonardo da Vinci's combining technology with art to learn about brain structure: One day he injected hot wax into the brain of an ox and thus generated a cast of the brain's ventricles, the hollow cavities that contain the brain's "transmission fluid."

Reeve told Groopman how his frustration at the snail's pace of science led him to establish his own research funding enterprise, the Christopher Reeve Paralysis Foundation, in 1999. A year before he died, Reeve said, "I heard a very distinguished scientist say, 'You can never do enough basic science.' These are the kinds of people I'm no longer willing to fund." Yet it is an inescapable fact that stem treatments, as they are developed, will benefit rats before they benefit us, at least in countries like the United States that require animal studies before human trials can commence.

In 2005, Hans Keirstead and his colleagues at the University of California at Irvine's Reeve-Irvine Research Center—named for Christopher Reeve—gave a preview of what the future may hold for people with spinal cord injuries, about a quarter of a million people in the United States alone. The scientists provided one of the first direct demonstrations that human embryonic stem cells can regenerate tissues damaged by spinal cord injuries in rats. Keirstead and his colleagues reprogrammed stem cells to create new myelin, the insulating sheath that protects nerve cells and allows them to communicate movement via electrical impulses. They then transplanted the stem cells into the site of the injuries, thereby restoring the ability of the paralyzed rats to walk, although the experiment was successful only in the group of rats that had been injured one week earlier and then treated, not in the group that had been injured ten months earlier and then treated.

Spinal cord injury patients, their families, and the entire stem cell research and therapy community took notice in January 2009 when the U.S. Food and Drug Administration granted clearance to Geron Corporation to begin the world's first human trial of embryonic stem cell–based therapy for acute spinal cord injury. Geron's cellular product, called GRNOPC1, contains nerve progenitor cells derived from human embryonic stem cells. These progenitor cells, called oligodendrocytes, show the ability to remyelinate nerve cells and stimulate nerve growth in laboratory animals. The phase 1 trial, designed to test the safety of GRNOPC1 in just ten patients, was put on hold by the FDA for more than a year after Geron reported an animal study in which a higher frequency of small cysts were observed within the injury site in the spinal cord of animals injected with GRNOPC1 than had previously been seen. An additional preclinical study established to the FDA's satisfaction that the cysts were within the normal range and the trial could resume, which it did in July 2010 with the first patient enrolled the following October. Geron is not alone in the cell therapy for spinal cord injury space. Early in 2011, the American firm StemCells, Inc. was given approval by Swiss regulatory authorities to begin treating twelve Swiss spinal cord injury patients with HuCNS-SC, its purified neural stem cell product.

The ability to create myelin through cell therapy would also be of potential benefit to people with multiple sclerosis, a debilitating autoimmune disease that affects the nervous system. In MS, the body's immune system mistakenly attacks the myelin around nerve cells. The scarring that results disrupts the nerve impulses that control sensation, strength, and coordination, causing symptoms from numbness and tingling of extremities to paralysis. The disease killed cellist Jacqueline du Pré and Texas congresswoman Barbara Jordan. Writer Joan Didion, actors Annette Funicello and Teri Garr, and Olympic medalist Betty Cuthbert have MS. MS affects more than a million people worldwide, including four hundred thousand Americans, and particularly young adults between the ages of twenty and forty. Twice as many women as men are diagnosed with MS. Multiple sclerosis is the third most common neurodegenerative disease after Alzheimer's and Parkinson's. Recently, scientists from Cambridge and Edinburgh universities in Britain found a molecular switch that helps stem cells regenerate myelin in the brains of laboratory rats. If the switch can be turned on with the right drug—and investigators think it can—MS patients may stand to benefit from enhanced tissue repairing activity.

Of all the neurological diseases that call for remedies and funding, none is as pressing from a public health perspective than Alzheimer's. For years experts have warned that as the U.S. population ages, especially the baby boom generation, Alzheimer's will become an enormous public health problem unless science finds a way to slow the progression of the disease or prevent it. A study of the disease based on 2000 U.S. Census data predicts that the prevalence of Alzheimer's will increase 27 percent by 2020, 70 percent by 2030, and nearly 300 percent by 2050.

The economic and social costs of the disease are large by any estimate. Today, annual costs are estimated to exceed $170 billion, including health care costs and costs associated with lost productivity. Disease spokespersons have stated repeatedly that Alzheimer's disease, which usually strikes after age sixty-five, threatens to destroy the health care system and bankrupt Medicare. A united front of prevention, intervention, and treatment offers the only hope of significantly easing the outlook posed by such alarming numbers.

About 5.3 million Americans suffer from Alzheimer's based on figures reported in 2010. What happened to President Ronald Reagan—what is happening to millions of people, mainly in industrialized countries—is the most massive assault on human memory the world has ever seen. The cause of Alzheimer's is still widely debated. What is generally agreed is that people with the disease have large deposits of a protein called beta amyloid that form plaques in their brain. A buildup of these plaques is believed to ultimately interfere with the transmission of neural messages in the brain and cause the death of brain cells involved in memory.

Internal representations of the external world eventually become encoded in brain protein molecules and stored. When a memory is recalled, based on recent studies, its coding structure takes on an active state. Then it is put once again into an inactive state, where it can be reactivated later by something we see, think, or experience. MIT neuroscientists, led by Nobel laureate Susumu Tonegawa, have found a molecular switch in the brain that controls the formation of lasting memories.

An army of scientists, entrepreneurs, and investors have banked their reputations, dreams, and capital on the possibility that the body has, within its power, the ability to turn its own stem cells into neural cells, thus achieving the perfect tissue match. The current strategy of salvaging functional neurons would be controlled by a more powerful force of replacing them. But ever since 1907, when German physician Alois Alzheimer first described the syndrome we know today as Alzheimer's disease, observing how it progresses from simple forgetfulness to mental decline and complete mental paralysis, this has been difficult to do.

Stem cells, for their part, are viewed as potential regenerators of brain tissue to replace brain cells lost as part of the Alzheimer's disease process. Yet the challenge is monumental, greater than the challenge of using stem cells to treat any other neurological disease. "Entire circuits underlying memory and thought, in many regions of the brain, are wiped out," wrote Sharon Begley in *The Wall Street Journal.* "Expecting transplanted neurons to weave themselves into the fraying circuits seems about as likely as a skein of yarn inserting itself into a damaged tapestry and recreating the original." The analogy is not unreasonable. Could stem cells already in the body be coaxed into cooperating?

The challenge facing scientists is to generate enough adult or embryonic stem cells or tissue for therapeutic replacement of damaged tissue in Alzheimer's disease. Adult stem cells can be directed to become a variety of specialized tissues with comparative ease, but they can be devilishly difficult to grow in large numbers. Embryonic stem cells can be directed down a desired pathway including toward the type of brain neurons that are affected in Alzheimer's disease. In 2011, researchers at Northwestern University reported progress in doing just that—generating a mostly pure population of functional basal forebrain cholinergic neurons, the types of cells involved in memory, from human embryonic stem cells. These laboratory-created brain neurons can serve as a platform for designing specific cell replacement therapies and drugs that can prevent neuron death, the hallmark of Alzheimer's disease. Generating a pure cell population in these kinds of cell differentiation experiments is vitally important. For cell replacement therapy, it is unlikely that the FDA will be willing to accept levels of accuracy in which one in a thousand cells, one in one hundred thousand cells, or even one in a million cells do the wrong thing, putting the person with neurological disease at unnecessary risk, especially when other treatments proven to be safe and effective may be available.

Perhaps no picture of the agony of families struggling with Alzheimer's disease is more telling than that of two California women: Nancy Reagan, widow of President Ronald Reagan, and Maria Shriver, wife of former California governor and movie star Arnold Schwarzenegger. President Reagan suffered from Alzheimer's for more than a decade. Just before his death in 2004, Nancy Reagan emerged as an outspoken advocate for embryonic stem cell research. Maria Shriver's book *What's Happening to Grandpa?* describes her experience with her father, Sargent Shriver—former Democratic politician, ambassador to France, Peace Corps founder, emeritus chair of the Special Olympics board—who suffered from early Alzheimer's disease for years before his death in 2011.

Just weeks before President Reagan's death, the former first lady appeared at a Hollywood gala to raise funds for stem cell research. She was introduced by fellow stem cell advocate, actor Michael J. Fox who went very public with his Parkinson's disease, a degenerative disorder of the central nervous system. Parkinson's is characterized by tremors, rigidity, and difficulty initiating movement. Its cause remains a mystery. Nationwide, it is estimated that more than five hundred thousand people suffer from the disease. The National Institute of Neurological Disorders and Stroke estimates that sixty thousand new cases are diagnosed each year in the United States, all at a cost of $6 billion annually for health care and lost productivity. The average age of onset is sixty years, but people as young as twenty develop the disease. Fox was diagnosed when he was thirty. The Michael J. Fox Foundation for Parkinson's Research has contributed more than $150 million to stem cell research in hopes of finding improved treatments and developing a cure for Parkinson's. Fox himself unleashed a political firestorm when he made television ads for Democratic candidates during the 2006 congressional elections. He was accused by conservatives, including radio talk show host Rush Limbaugh, of exaggerating his tremors for partisan political purposes.

In the view of some, Parkinson's disease is the leading candidate for stem cell success of all neurodegenerative diseases. The challenges may be more manageable than, say, rewiring the neural-memory networks that disappear in patients with Alzheimer's. In Parkinson's, you may not need to rewire nerves so much as get stem cells into the brain so they can restore function to the substantia nigra. The substantia nigra produces dopamine, a chemical messenger that transmits signals to the parts of the brain that control and coordinate movement. As cells in the substantia nigra begin to die, the amount of dopamine produced declines. That brings on the characteristic tremors of Parkinson's. Stem cells that have been programmed to grow into substantia nigra cells or cells like them could be implanted into the brain, where they would release dopamine in the customary manner and provide relief from the debilitating consequences of the disease.

When embryonic midbrain neurons are transplanted into patients with Parkinson's, some of these patients respond well. These transplanted cells are not stem cells, but immature neurons from the region of the substantia nigra. In 2004, investigators from the Sloan-Kettering Institute and Cornell University in New York and the University of Connecticut succeeded in the laboratory in using human embryonic stem cells to make brain neurons that secreted dopamine. Yale University researchers subsequently took endometrial stem cells from the linings of rat uteruses and showed that they can become dopamine-secreting cells when transplanted into the substantia nigra. Because endometrial stem cells are readily available and easy to collect, the National Institutes of Health suggested that "banks of endometrial stem cells could be stored for men and women with Parkinson's disease." It is yet another example of the malleability of stem cells when placed in a different environment, in this case an environment that can direct their differentiation down a therapeutic pathway.

Just a decade ago, the typical neurology textbook taught that neurons in the central nervous system could not regenerate. That teaching has been eclipsed by discoveries in stem cell biology. The hope now is that stem cells may be able to regenerate cells to repair damaged nerve tissue. That hope is the basis for a human clinical trial launched in 2010 by the British firm ReNeuron for patients with stroke, an interruption of the blood supply in any part of the brain. Stroke is the leading cause of long-term disability in the United States, afflicting nearly 800,000 people and claiming more than 170,000 lives each year. Patients in the trial received ReNeuron's neural stem cell treatment, ReN001. By injecting these cells into patients' brains, the hope is the cells will repair areas damaged by stroke and improve both mental and physical function, as they do for laboratory animals. Therapeutic transplants of stem cells are one of the best long-term hopes for reversing progressive neurological disease and the debilitating effects of stroke. Taken as a whole, the hope for treatments of neurodegenerative diseases and disorders that stem cells may offer has energized families and patient advocacy groups to push hard for federal funding for research and favorable stem cell research policies.

UNIVERSAL DONOR CELLS

Stem cells have the potential to become the "universal donor cells" long sought by scientific medicine. They could become the cells that would treat a wide array of acute, chronic, and genetic diseases and debilitating conditions. They could transform medicine and be an unprecedented humanitarian benefit. The reprogrammed skin cells developed by laboratories in Japan and Wisconsin that captured the public's imagination in November 2007 may have a bright future, but not as universal donor cells. These cells would be specifically tailored for individual patients. Even if skin cells or other mature adult cells are successfully engineered in the laboratory to become stem cells and are proven to be safe and effective in patients, the economics of individualized patient-specific treatment—usually termed personalized medicine—will likely not work out for stem cell therapy. Manufacturing of autologous cell therapies—therapies in which a patient's own cells are processed and purified for treating the same patient—is much more complex than manufacturing one distinct cell type for use in many patients: allogeneic cell therapy, which acts more like a drug that can potentially be used by many people despite their different genetic makeup. That may change with advances in technology, but right now, allogeneic cell therapy is a more cost-effective approach.

The list of challenges to stem cells' becoming universal donor cells for transplantation is a long one. Will it be possible to reprogram adult stem cells found in bone marrow and other tissues so that they can participate across the board in damaged tissue repair and restoration of function lost to disease or injury? Will it be possible to program embryonic stem cells to achieve the same goals by directing their differentiation into the type of cell needed for therapy? Will it be possible to use cells foreign to a person to treat that person without triggering an overwhelming immune system rejection? That is, would those stem cells be intrinsically compatible with a person's immune system or can they be bioengineered to be compatible? Is it possible that unique stem cells from umbilical cord blood, which lack the immunological markers that would make them targets for rejection if transplanted, have the ability to grow bone, cartilage, neurons, liver, and heart muscle? Might these cells be a universal source of stem cells for tissue repair and tissue regeneration? We won't know for some time. Meanwhile, the effort will engage scores of scientists around the world for decades to come.

The history of transplantation medicine, a field that began in earnest in the mid-twentieth century, has faced the challenge of tissue compatibility since its inception. Regenerative medicine using foreign cells will be no different. An understanding of how to prevent rejection of transplanted stem cells is what a National Academy of Sciences panel described as "fundamental to their becoming useful for regenerative medicine." Without conquering that barrier—a barrier now under massive assault by both basic science and applied technology—the future of stem cells in medicine could be limited. But immune rejection is only one of many challenges. Improved methods for transplanting embryonic stem cells will need to be developed. The cells will also need to be monitored closely to make sure that their chromosomes are stable prior to transplantation. The chromosomes of cells in cell lines grown in the laboratory, especially embryonic stem cells and induced pluripotent stem cells (reprogrammed adult cells), tend to undergo rearrangement, multiple cell divisions, and have highly variable levels of gene expression in culture. Such gene "shuffling" and variable gene expression could present a real danger to humans, as could the tendency of these stem cells to form benign tumors called *teratomas* as they do when transplanted into laboratory animals. The formation of teratomas in these animals is actually proof of the pluripotency of embryonic and reprogrammed adult cells, as we will explore in the next chapter. The challenges are to direct these pluripotent cells to produce only a desirable type of cell—and in pure populations—that possesses repair capabilities for a specific tissue, and to be able to deliver enough of these pure differentiated cell populations to the place in the body where they are needed.

Some contend that it has been disingenuous on the part of advocates of embryonic stem cell research not to acknowledge more clearly the potential risks associated with putting these cells into patients. If even a small percentage of patients develop tumors following transplantation of embryonic stem cells, how could their use be justified to treat non-life-threatening disease, especially when a treatment is already available, as in diabetes with current insulin therapy or in Parkinson's disease with "deep brain stimulation" medical devices? Advocates must recognize the consequences of that risk. Would you accept that risk for your child? Will the FDA?

However these questions are answered, it is indisputable in the eyes of many if not most biomedical scientists that human embryonic stem cells hold a vast reservoir of knowledge about disease, as well as human development and evolution. Scientists will continue to search for ways to recruit adult stem cells— those stem cells already at work in our tissues—more fully to the task of repairing these tissues when they break down in disease. At present, adult stem cells appear to be willing to serve such a role in some diseases, but not in others. They are not, at least not yet, the universal donor cell for transplantation.

Yet even if universal donor cells prove to be elusive as a therapy, stem cells have an unrivaled future for drug development. Much preliminary testing of new drugs is done using model cells such as fibroblasts, cells that give rise to connective tissue. These cells yield important information about a new drug's potential toxicity to our liver, for example, but they are a poor substitute and do not measure particularly well the effectiveness of the new drug on the cause of a problem in a given disease. Liver cells, called hepatocytes, cannot be grown in the laboratory unless they are genetically altered. That makes using hepatocytes for drug metabolism and toxicity testing or researching liver disorders a big challenge. That challenge may be on the way to being met. In 2010 researchers at the University of Cambridge made induced pluripotent stem cells from the skin of patients with liver disease and then created hepatocytes from the reprogrammed cells. Hepatocytes obtained from patients' iPS cells could serve "as a platform for drug hepatotoxicity assays and to individualize patient therapies," wrote Linda Greenbaum of the Thomas Jefferson University School of Medicine in a commentary accompanying the published study.

As scientists figure out how to coax stem cells to make the target cell type—a specific white blood cell, a skeletal muscle cell, a heart cell—the biotechnology and pharmaceutical industries are poised to bring their enormous R & D clout to bear. They will develop innovative drug therapies by testing them on these differentiated stem cells. That would usher in an exciting new era for drug development, joining the malleable powers of the stem cell with the "lock and key" biochemistry of the body.

We have reason to hope that stem cell therapies will relieve suffering and restore health to people with debilitating and catastrophic diseases. That hope is drawing ever closer to becoming reality as geneticists, cellular and molecular biologists, reproductive scientists, and bioengineers around the globe help solve the mystery of stem cells. What once seemed the stuff of science fiction has become a fantastic voyage into the very real world of stem cell research. A world where the promise of stem cell therapies will be fulfilled if scientists, politicians, ethicists, religious leaders, and ordinary citizens can agree on policies governing such research. A world that nonscientists must comprehend, at least on a basic level, so that all can participate in the great stem cell debate.

Chapter Two

ARCHITECTS OF DEVELOPMENT

Although nature commences with reason and ends in experience,
it is necessary for us to do the opposite—that is, to commence with
experience and from this to proceed to investigate the reason.
—Leonardo da Vinci

For centuries, the world was imagined as a harmonious whole known as the Great Chain of Being, a chain that joined the spiritual and natural worlds, heaven and earth. The Renaissance carried the idea of the harmonious whole into explorations of the human body. For the first time, the body was seen as a microcosm of the universe. To fully understand the world around us, we needed to explore the world within. We needed to know what we are made of and how we work.

No one gave more life to that idea than Leonardo da Vinci. In his anatomical drawings, Leonardo used his pen as a scalpel to lay open the surface of the body, to show its internal structure, to transform dead elements into a living, ordered whole. His famed drawing of *The Vitruvian Man* demonstrates just how fully

Leonardo viewed the human body as a harmonious whole of mathematical proportions. The Vitruvian Man's navel, the exact center of his body, is the motionless point where the artistic and scientific individuality cultivated by the Renaissance had its symbolic origin—the timeless pivot of life's compass.

The human body develops and functions at a level far more intricate than Leonardo had time and tools to discover. The body is designed and governed by cells, and the cell most responsible for maintaining harmony within the body is the stem cell. The stem cell is an unrivaled creative agent of growth and development in the Great Chain of Being. It is the architect and engineer of all complex organisms. Stem cells not only make crucial decisions for the developing body, they also play a vital regulatory role in homeostasis, which is the body's built-in ability to keep its internal systems in a state of biochemical balance—in harmony.

How do stem cells maintain our internal harmony? They create and regulate themselves through a continual process of self-renewal, self-differentiation (or specialization), and self-destruction through programmed cell death for unneeded or damaged stem cells in the body. The secret to their ability to both reproduce themselves and create new tissues is their "duality of purpose": The stem cell can copy itself exactly *and* yield specialized progeny. Stem cells of the very early embryo typically produce identical daughter cells rather than self-copies because they must build tissues rapidly. Then later on they assume the "duality of purpose" capability, which maintains a supply of stem cells to assist in tissue repair and homeostasis. Stem cells take cues from the world around them to modify their own genetic program and influence that of their offspring. The neighborhood or "microenvironment" in which they reside, sometimes called the stem cell *niche*, is crucial to their function. Nothing is more fundamental to understanding how stem cells work than this.

The discovery that our bodies have cells that possess even the vestiges of self-renewal has reawakened age-old dreams of regeneration—of living forever, or at least living longer. Dreams of regeneration and immortality have flowed deeply through our ideas of life and death, faith and progress, and national and cultural myths since the beginning of recorded history. Take the *Epic of Gilgamesh*, the oldest story known. Sometime around 3500 BC in ancient Babylon, King Gilgamesh of the Uruks journeyed through the twelve leagues of darkness in a desperate quest to find a plant that could restore youth. Gilgamesh finds the plant only to have it snatched from him by a snake, and his pursuit of immortality is for naught.

It is another ancient myth, however, that lends itself best to the story of regeneration: the story of Prometheus, son of the Greek god Titan. Prometheus steals fire from the gods and gives it to humans so that they may be enlightened by civilization and the arts. To punish Prometheus, Zeus has him chained to the side of Mount Caucasus, where a bird of prey feeds every day on his liver. The organ regenerates itself as quickly as it is devoured.

If the bird had fed on another organ—say, the heart or the brain—Prometheus would not have survived. That's because the heart and brain do not naturally regenerate themselves like the liver or skin. But that neat dichotomy is no longer so neat. Advances in our understanding of cell biology, embryo development, and genetics are causing scientists to rethink regeneration. The revolutionary fact they have discovered is that with each step in the development of our organs—from their earliest appearance in the embryo to their maturation upon birth—seeds of regeneration are left in various tissues and organs. Those seeds are stem cells.

STEM CELLS 101

The regenerative possibilities of stem cells have given the Promethean myth top billing today, Nadia Rosenthal wrote in *The New England Journal of Medicine:*

> We mere mortals do not possess livers with quite so vigorous a regenerative capacity, but the legend captures well the remarkable potential of the body to rebuild itself. Throughout our lives we sustain less gruesome injuries from which we recover spontaneously, often without realizing we were hurt. Wound healing involves the recruitment and proliferation of cells capable of restoring tissues and even organs to their original form and function. These cells must retain a collective memory of the complex developmental process by which the tissue was first constructed.

There are two basic categories of stem cells: adult stem cells and embryonic stem cells. The stem cells with the "collective memory" that Rosenthal mentions are *adult stem cells,* or mature stem cells, that are present in the organs and tissues of animals and humans after birth. The adult stem cell's ability to regenerate tissue comes from its ability to "remember" how the tissue was created in the first place. Creating that tissue is the role of the *embryonic stem cell,* which is found only in embryos.

All stem cells can renew themselves and develop into specialized cells. For many adult stem cells the potential to regenerate is limited, more or less, to the tissues and organs in which it resides. In other words, an adult stem cell from the liver regenerates liver tissue. But some adult stem cells are more versatile; they have more "plasticity" in regenerating tissue types. Such adult stem cells are known as *multipotent* for their potential to regenerate beyond the tissue in which they reside. An adult stem cell from the liver, for example, can be coaxed or reprogrammed to regenerate tissue for the kidney. Multipotent adult stem cells are also found in umbilical cord blood, amniotic fluid, and bone marrow and have the ability to repair and regenerate many types of tissue. The embryonic stem cell gives rise to all of the tissue cell types in the body, more than 220 types. The embryonic stem cell is *pluripotent* because it represents a type of cell that builds entire bodies—for mammals in a matter of months. There are some emerging studies that show that certain adult cells from bone marrow may have all the capabilities of embryonic stem cells, but this remains to be seen. In any event, based on the consensus of scientific opinion to date, adult stem cells do not build bodies; embryonic stem cells do.

Adult stem cells exist in small numbers in tissues that constantly renew themselves such as blood, skin, adipose or fatty tissue, the epithelial lining of various organs, placenta, testes, and others. Mesenchymal stem cells, a type of adult stem cell, are found in many tissues, can be grown in the laboratory, and have the ability to form bone, cartilage, muscle, and fatty tissue. But adult stem cells are typically tissue-specific, residing in a given tissue and regenerating only that tissue or related tissues. For the purposes of research, adult stem cells are usually obtained from bone marrow, from blood in the umbilical cords of newborns, and sometimes from circulating blood. The stem cells are extracted from the blood and marrow using a specialized centrifuge. Adult stem cell lines are difficult to grow in the laboratory and generally die off after some number of replications. A lot of stem cell research is performed using adult stem cell lines from laboratory animals such as mice and rats because obtaining adult stem cells from humans other than from their blood can involve tissue biopsies, which are by nature invasive or surgical procedures.

Embryonic stem cells, which grow readily and infinitely in the laboratory, come from embryos that are created at fertility clinics but not used. As developmental biologist Jonathan Slack has noted, technically, the embryonic stem cell "does not exist in nature." It is a product of laboratory tissue culture. The leftover embryos from which embryonic stem cells are derived are typically no older than fourteen days. They are donated by fertility clinics to research facilities with the permission of their "owners." From leftover early embryos called *blastocysts*, scientists can extract cells from the inner cell mass and make stem cell lines. These new "families" of stem cells each contain the same DNA. They are "clones." Embryonic stem cells can also be made in the laboratory using a process called *somatic cell nuclear transfer*, or SCNT.

In SCNT, the DNA from a body cell of a donor is put into an egg from which the DNA has been removed. The SCNT technique produces embryonic stem cells that are genetically identical to the donor's cells. The process is often referred to as *cloning*, with variations on the theme. In reproductive cloning, a genetically identical organism is created through the SCNT technique, as was famously demonstrated in 1997 with Dolly the sheep. In therapeutic cloning, the stem cells are created through SCNT that can potentially be used to treat the specific illness of a specific person who donated the DNA. In research cloning, largely an interchangeable term for therapeutic cloning, the stem cells are used to advance the study of science and medicine. In each case, a *blastocyst*, the early form of the embryo consisting of a clump of about two hundred cells, is the source of the stem cells.

Again, nuclear transfer in this context involves transplanting the nucleus from any cell in the body *other* than a reproductive cell into an egg whose original nucleus has been removed. They are fused with the aid of an electric pulse. As the new cell divides, it results in a new embryo. In cloning, the SCNT process results in the creation of an embryo without using sperm. It does not involve sexual reproduction. In the fertility treatment known as *in vitro fertilization* or IVF, embryos are created using both eggs and sperm. IVF does involve sexual reproduction, albeit in a petri dish. Both SCNT and IVF are laboratory techniques that can produce embryonic stem cells, but SCNT could produce stem cells that would be a genetic match for a patient, something yet to be accomplished in people. Shoukhrat Mitalipov and his research team at the Oregon National Primate Research Center accomplished the feat in monkeys, cloning them and growing genetically matching stem cell lines from the clones. Scientists had been trying to clone primates for many years, without success. Science is replete with techniques and treatments that may work in mice or rats but fail to work in monkeys or human beings. But this time it did work in monkeys. Mitalipov and his team published a landmark paper in *Nature* in 2007 describing how they used SCNT to make rhesus macaque embryos and then created embryonic stem cell lines from two of the embryos. The DNA of the embryonic stem cells was a genetic match with the DNA of the donor macaques. Macaque eggs used in the experiment reprogrammed the genomes of the donated nuclei. Then the cloned blastocysts began dividing and making embryonic stem cells. "I am quite sure it will work in humans," Mitalipov told reporters. Four years later, the successful cloning of a human blastocyst had yet to be reported in the scientific literature, owing to the difficulty and expense of SCNT experiments and the inability of scientists to find adequate supplies of human eggs.

It is the manipulation of stem cells that lies at the heart of the debate of the twenty-first-century biorenaissance. Creating embryonic-like stem cells and reprogramming adult stem cells to do something other than what they're originally programmed to do makes possible the practice of regenerative medicine—the ability to repair and restore tissues, organs, even limbs damaged by disease or lost to injury. Questions about regeneration were first raised by scientists who studied the amazing power of flatworms, newts, and other lower animal forms to replace or restore tissue they had lost to injury. Scientists asked back then, as countless other researchers have asked since, "If worms can regenerate themselves, why can't we?"

Thanks to stem cells, maybe we can.

A BRIEF HISTORY OF REGENERATIVE SCIENCE

Given all the news about stem cells and regeneration, you might think the field is fairly new. Not so. Practically from the beginning of the organized study of biology, scientists have aided and abetted the cultural quest for greater longevity by studying lower animal forms in an effort to close what has been called the "regeneration gap" between worms, salamanders, and fish on the one side and humans on the other.

Animal regeneration science began almost three hundred years ago with René-Antoine Ferchault de Réaumur, who studied the extraordinary ability of crayfish to regenerate their limbs and claws that were damaged or removed. But the field really took off in 1740 with Abraham Trembley's studies of the *hydra*, the pond-dwelling polyp named for the many-headed water serpent of Greek mythology that regrew a new head for every one that was cut off. When Trembley, a Swiss scientist and mathematician, snipped a hydra in two and observed it overtime, he noticed how the pieces eventually re-formed as separate entities, each with a head and a full set of tentacles. How was that possible? It took another 150 years of science for the question to be answered, at least in part.

The history of regenerative science involves three closely related branches of science: cell biology, embryology, and genetics. It is through cells that tissue is developed, regenerated, and repaired. It is through cells that species are perpetuated, altered, and equipped with new features, thanks to DNA—the genetic information of life that resides in the nucleus of the cell. In mammals, cells fall into two broad classes. *Somatic cells*, a.k.a. body cells, consist of all the cell types that develop our various organs. *Germ cells*, a.k.a. reproductive cells, consist of only those cells that allow reproduction of the species. Germ line cells in the testes and ovaries make sperm and eggs. They are immortal in the sense that they generate a whole organism after fertilization and then make the sperm or eggs of that organism, linking one generation to the next. Stem cells are found in both of these broad classes of cells, somatic and germ cells.

The term "stem cell" debuted in the 1890s with the rise of modern embryology, the study of the embryo and its development. German zoologist Valentin Häcker first used the word *Stammzelle* in 1895 to describe the cell in the early embryo of the crustacean *Cyclops* that gives rise to primordial germ cells—cells that eventually make eggs and sperm. But it was Edmund Beecher Wilson, an invertebrate zoologist at Columbia University, who gave legs to the term "stem cell" in his 1896 book *The Cell in Development and Inheritance.*

Wilson saw how the primordial reproductive cell differs from the body cell in roundworms, "not only in the greater size and richness of chromatin of its nuclei, but also in its mode of mitosis [cell division]." Chromatin is the material inside the nucleus of a cell that makes up chromosomes. The x-like bodies in the nucleus of the cell, chromosomes had recently been discovered by the German biologist Walther Flemming who used synthetic dyes to stain cells and reveal chromosomes in their nuclei. Zoologist Theodor Boveri, Flemming's countryman, traced the division of germ cells all the way back to the two-cell stage in the blastocyst, the earliest stage of the embryo. Wilson correctly observed that a blastocyst produced two kinds of daughter cells. One daughter cell is destined to become a body cell, he wrote, while in the other, "which may be called the *stem-cell*, all the chromatin is preserved and the chromosomes do not segment into smaller pieces." He then made a prophetic connection between chromosomes and DNA: "that inheritance may, perhaps, be effected by the physical transmission of a particular compound from parent to offspring." As Stephen Hawking put it, "If DNA is the software of genetics, then stem cells are its hardware."

With Wilson's work, the fields of cell biology, embryonic development, and genetics were brought together for the first time. The affiliation didn't last long, however, and the study of embryology and genetics soon went their separate ways. No one was more responsible for creating those separate paths than Wilson's next-door colleague at Columbia University, Thomas Hunt Morgan. Ironically, it was Morgan, probably more than anyone else, who set the stage for stem cells to bring genetics and embryology back together again at the end of the twentieth century and help guide the genetic revolution we are just beginning.

If Gregor Mendel is the father of genetics, Thomas Morgan is his firstborn son. And if the room at Cambridge University's Cavendish Laboratory, where James Watson and Francis Crick put together their model of the structure of DNA, is the room of rooms in the history of the life sciences, Morgan's "Fly Room" at

Columbia is a close second. Eight desks and a kitchen table were stuffed into the 621-square-foot space in Uptown Manhattan. A powerful smell of fermenting bananas pervaded the room. The bananas were for feeding, breeding, and otherwise making happy *Drosophila,* the fruit flies that empowered genetics and made Morgan world-renowned.

The flies were Morgan's ticket to Stockholm in 1933 for the Nobel Prize, which he was awarded because he proved that the chromosomes carry the genetic information and pass it on through successive generations of cells and through successive generations of organisms. Morgan brought to life Mendel's laws of inheritance by focusing on how genes are transmitted through generations of fruit flies and how mutations can yield genetic variations. In other words, Morgan was able to explain how a male fruit fly was born with white eyes instead of the red eyes of his siblings.

Genetic research early in the twentieth century prepared the soil for the wave of hybrid seed varieties that swept over American cropland in the 1930s. It was also applied to animal breeding. It was, in many ways, the first genetic revolution, says Garland Allen, Morgan's biographer. It brought together Mendel's field research on heredity with experimental laboratory science like Morgan's in the broader context of industrial growth. Genetics was where the future was going, not embryology.

Yet Morgan's earlier work with the *planarian* flatworm heralded the future of regenerative science when he made an important observation about the flatworm's regenerative properties. "When an animal reaches a size that is characteristic for the species," he wrote in his 1901 classic *Regeneration,* "it ceases to grow, and it may appear that this happens because the cells of the body have lost the power of further growth. That the cessation of growth is not due to such a loss of power is shown by the ability of many animals to regenerate a lost part." Charles Darwin himself had observed the amazing regenerative powers of *planaria* in Brazil during his famous voyage on the HMS *Beagle*. "Having cut one of them transversely into two nearly equal parts, in the course of a fortnight both had the shape of perfect animals," Darwin wrote in his journal. Though it was a well-known experiment, he found it "interesting to watch the gradual production of every essential organ" in the half of the transected worm that originally lacked these organs.

Researchers who followed in Morgan's footsteps learned that the flatworm's secret of regeneration was in its ability to mobilize populations of regenerating stem cells at the site of injury. Subsequently, while studying newts and zebra fish, scientists discovered that certain stem cells can travel back in time, assume a more potent state, then reorganize and rebuild the lost limb from the ground up. These more potent stem cells had the potential to do something other than what they were destined to do: they could switch allegiance from one cell type to another to make any tissue in the body. Scientists now believe that this power to change destiny and function is key to understanding the potential of adult stem cells in humans. That power is called *pluripotency*, and Morgan's contemporary successors are intent on bringing it forth in full.

CLOSING THE "REGENERATION GAP"

One of Thomas Hunt Morgan's present-day successors, Elly Tanaka of the Max Planck Institute in Dresden, Germany, says the future of regeneration for humans hinges on at least three biological principles: understanding the role of a mound of cells called the *blastema* in the ability of lower animals to regenerate whole limbs or parts of limbs from a stump; understanding the role of pluripotency—that is, how a cell switches its fate from being locked into performing a specific task to doing something else for a living; and understanding how cellular memory works to produce gradients, the appropriate spatial order and pattern for the formation of a new limb, a process called *morphogenesis*.

If we humans are ever to possess the ability to regrow limbs lost to our forerunners through the eons of our evolution, we probably are going to have to know a lot more than we do about the blastema. The regenerating blastema is a pool of immature cells that forms over and beneath the site of a severed limb or tail in animals like salamanders and newts. It is the underlying tissue of the blastema that calls the shots about what should be regenerated. That underlying tissue is a complex mix of cell types including skin cells, cartilage cells, and nerve cells. Regenerating blastema cells can become the cell type they need to be at a given time. Muscle cells, for instance, can contribute to regenerating cartilage; nerve cells can contribute to building muscle. Blastema cells are probably best described as heterogeneous, meaning they are a mixture of progenitor cells that can form a variety of different tissues including muscle, cartilage, and neural tissue that insulates the nerves of the limbs.

If there is a single biological capacity that separates a salamander from humans—aside from our ability to reason in most cases—it is the capacity of salamander cells to regress to an earlier, more potent state when injured and become cells with greater regenerative power to form replacement tissue. Are salamander cells "intrinsically different" from human cells or those of other mammals because they respond to injury signals in a unique way? The answer isn't in yet. What is known is that salamander cells switch their fates much easier than human cells. One possible explanation is that the blood of salamanders possesses a special protein or growth factor that stimulates the cell-fate decision of adult cells by manipulating the cell cycle. Another idea is that special signaling proteins break down the muscle tissue, causing cells to lose their ability to make muscle protein, and converts them into the type of adult stem cell typically found in blood and bone marrow. Finding these proteins and the genes that regulate them will help unveil the mystery of new limb formation.

Because of dramatic advances in biochemistry, computer software, and robotics, it is possible to screen vast libraries of genes and chemical compounds to find those that influence cell behavior in particular ways. By screening an array of small synthetic molecules, scientists at the Scripps Institute and the Genomics Institute of the Novartis Research Foundation in La Jolla, California, found one that causes adult stem cells to undergo reverse cell specialization as they do in the blastema that forms after a salamander loses a limb. When treated with the molecule, muscle cells move backward from their normal identity into earlier forms of stemlike cells that don't make muscle proteins. The molecule produced an effect similar to that of the mouse *MSX1* gene, throwing a mature specialized cell into reverse. The scientists named the synthetic compound "reversine" and described it as one of the first steps in mimicking the natural regeneration found in salamanders. After discovering reversine, Sheng Ding and his Scripps team found another molecule they named "pluripotin" that keeps embryonic stem cells in a pluripotent state and other organic compounds that direct embryonic stem cells in mice to become heart cells and neurons. Drugs like reversine and pluripotin "are the future of regenerative medicine," Ding said, suggesting it is likely that drugs will be discovered that will cause cells, tissues, and organs to organize self-repair in animals and in us. Ding's words proved prophetic when he and his colleagues used a cocktail of small drug-like molecules (instead of genes, which are ferried into the cell's genome using potentially cancer-causing viruses) to reprogram adult cells into a pluripotent state.

Another mystery is how the blastema "knows" what structures to make and how to limit its energies to making only the missing parts. "Do the cells have access to global information about what structures are present in the limb?" asks developmental biologist Lewis Wolpert in *The Triumph of the Embryo*. If so, how did they get it? "The evidence is that they do not have such information and their behavior is totally determined by very local events at the cut surface," Wolpert said. In regeneration, the cell acts locally to bring about a global solution, limb restoration. It takes stock of its new geography at the outer perimeter, not the geography conferred on it during embryonic development. It has a new position in the global order of the limb—not where it actually is at the surface of the break, but at the far boundary, drawing the regenerative machinery ever outward until the new limb is complete. The blastema cell changes the expression of its genes—just as the egg cell with genes imported from a body cell reprograms the genes to make a cloned mammal like Dolly the sheep.

Wolpert describes how the French flag helps to illustrate the idea of biological gradients and the challenge for regenerating tissue: "Given that a line of cells that can be blue, red, or white, how should they communicate with each other so as to form a French flag that is one-third red, one-third white, and one-third blue, and continue to do so even when parts are removed?" The answer lay in the ability of cells in regenerating systems to "know" where they are in a coordinate system, a gradient, to possess positional information. It was British mathematician and computer genius Alan

Turing who put us on the path to understanding the chemistry of growth in biological gradients, that is, in complex organisms like us. He pioneered critical features of our current understanding of form in development, or morphogenesis.

How the blastema cells acquire what is called the positional information they come to know—their anatomical address in a cellular map or gradient—and then act on it is something else we will need to find out about if we want to make limb regeneration our own. Developmental biologists studying model organisms that regenerate limbs have been asking questions about these amazing feats for a long time. But a new breed of scientist is introducing new ideas, new tools, and new energy to the naturalist's approach of the past. These new scientists are steeped in molecular biology and getting acquainted with the emerging field of bioinformatics, which brings powerful supercomputers and statistical methods to bear on the analysis of biological data, particular stem cell biology.

One of the new breed of molecular scientists is David Gardiner of the University of California at Irvine, who thinks that focusing on the first rather than the final stage of limb regeneration in salamanders is the way to go. Historically, scientists have focused on the final stage—the reappearance of missing structures and respecialization of cells—rather than on wound healing and specialization. Gardiner has a different view, according to science writer Brian Alexander:

> Gardiner's view is that regeneration happens in three stages. Because of the limitations of past technology, scientists spent most of their time looking at stage 3, after the process is well along. To Gardiner, this is like trying to jump the Mississippi at New Orleans. He'd prefer to jump the small stream way up in Minnesota, at the very beginning of stage 1.

To prove his point, Gardiner is building an "infrastructure of gene sequences," a library of salamander genes. With these references in hand, he and his colleagues hope to tease out the genetic signals that begin the earliest steps in regeneration.

Genes account for the "stemness" of pluripotent cells, but so does their environment—their interactions with other cells within the "neighborhood" or niche where they reside. "In [newts and zebrafish], researchers are now investigating where the cells involved get their instructions from and which genes and proteins are responsible for regeneration," wrote Helen Pearson in *Nature*. Genes such as the *Notch* stem cell gene, which makes cell membrane proteins that may be important in disorders affecting memory like Alzheimer's disease. Genes such as the master gene *nanog*, which makes proteins that preserve stem cell self-identity and self-renewal during the growth of the embryo. Discovered in 2003 by scientists in Scotland and Japan, *nanog* was named after Tir Nan Og. *Nanog* is a member of an ancient family of genes called *homeobox* that executes the three-dimensional body plan early in embryonic development. *Homeobox* genes are key to morphogenesis in animals and plants and limb regeneration in salamanders.

Armed with that kind of genetic control information, "you could create stem cells when you need them, where you want them," said Mark Keating, a Harvard University cell biologist and cofounder of Hydra Biosciences, a Massachusetts-based start-up aiming to pioneer molecular regeneration and pain management medicines. His discovery of a gene called *MSX1* that can cause mature mouse muscle cells to become "cells that are like stem cells" is a case in point. Other genes with names like *Oct-4*, *BMP*, and *SOX2* also make proteins that play an important role in pluripotency and regeneration. Sorting out the molecular and genetic basis for regeneration "may well usher in a golden era for regenerative medicine," wrote Shannon Odelberg, a member of Keating's research team.

That era may already be at hand. The story begins in July 2006 when Japanese regeneration scientist Shinya Yamanaka stunned attendees of an international stem cell meeting with a report that he had used a recipe of four genes, including *Oct-4* and *SOX2*, to equip skin cells from mice with embryonic stem cell-like powers. *Oct-4* and *SOX2* make *transcription factors*, powerful proteins that bind to DNA and control when genes are switched on or off and thus the level of gene expression. Yamanaka had already made a name for himself as the leader of the Japanese research team that codiscovered the master stem cell gene *nanog*. One of the attendees wondered whether Yamanaka had "hit a home run." Suddenly, it seemed that reprogramming adult cells to a pluripotent state might not be decades into the future or even the stuff of science fiction. A year later, Rudolf Jaenisch and his colleagues at MIT and Harvard shared the media spotlight with Yamanaka's team and another team when the three research groups confirmed and extended Yamanaka's original experiments. The scientists tagged the reprogrammed cells with fluorescent markers and injected them into mouse embryos. The live mice from those embryos showed that the reprogrammed cells glowed with fluorescence dye marker and contributed to the formation of all their tissues. Subsequent generations of mice also carried the markers, proving that the reprogrammed cells had entered the germ line of the mouse strain. That prompted Yamanaka to say that he expected his technique to yield a "big success" — reprogramming human skin cells to behave like embryonic stem cells.

Just six months later, in November 2007, Yamanaka published the "big success" he had predicted. It was a feat shared by his laboratory and that of University of Wisconsin stem cell pioneer

James Thomson. With the Japanese researchers writing in *Cell* and the Wisconsin researchers writing in *Science*, the two teams described the methods for transforming skin cells into embryonic-like stem cells, called induced pluripotent stem cells. Both groups used just four genes, including *Oct-4* and *SOX2*, to reprogram and dedifferentiate the cells, taking them back to a largely undifferentiated state. Once there, the cells were tested to see if they really possessed the characteristics of embryonic stem cells. The scientists found that the reprogrammed cells yielded clumps of embryonic tissue called embryoid bodies when suspended and stimulated in cell culture. Upon analysis, these embryoid bodies were found to represent all three embryonic germ layers of cells that build the human body and its 220 tissues. (See the section "Gastrulation and a Womb with a View.") Both the Japanese and Wisconsin teams used viruses to inject the genes into the skin cells and transform them. Problems with using viruses to insert genes into chromosomes have dogged the field of gene therapy for many years, but scientists expressed confidence that this challenge would be overcome by using chemical compounds to assist in carrying out the transformation. As we saw earlier in this chapter, Sheng Ding's laboratory proved them right when Ding and his colleagues used a druglike cocktail of small molecules to reprogram the cells. Harvard researchers subsequently reprogrammed different types of adult cells to pluripotency using synthetic RNA molecules, and they did so at a level efficiency that had not been previously reported. Efficient reprogramming is necessary before these cells can be used in therapies. The creation of pluripotent stem cells from mature adult cells was a technical tour de force unlike anything the field had seen since Thomson isolated human embryonic stem cells nine years earlier. Today, stem cell laboratories around the world are creating induced pluripotent stem cells at will and discovering just how malleable the differentiation program of the cell actually is. Yamanaka had indeed hit a home run—a grand slam!

THE GENETICS OF STEM CELLS

It was the discovery of so-called pluripotent stem cells in mice that launched the field of modern stem cell research a half century ago. In the 1950s, at the Jackson Laboratory in Maine, developmental biologist Leroy Stevens conducted experiments on sex cells in reproduction, especially reproductive tissues in mice. Stevens discovered an unusual growth in a strain of mice: a tumor of the testes called a teratoma. Using some nifty detective work over a decade, Stevens tracked the origin of the teratoma to the cells that make sperm. When the teratomas were transplanted into mouse bellies, the tumors tried to organize into actual embryos. Stevens further discovered that individual cells from the teratomas were able to develop into a variety of tissues. Stevens called these embryonic stem cells "pluripotent," and the field that is revolutionizing biomedical and reproductive science today was born.

About 10 percent of *planarian* flatworm cells retain their pluripotency, even in mature adults. Human embryonic stem cells appear to lose their pluripotency about three weeks following fertilization and early embryonic development. "One of the objectives of our research is to understand the mechanisms underlying *planaria's* ability to retain pluripotent stem cells," said Kiyokazu Agata of the RIKEN Center for Developmental Biology in Japan. "There must be a set of genes responsible for the continued presence of pluripotent stem cells." So far, Agata and his colleagues have found more than one hundred genes that are expressed in the stem cells. They are now in the hunt for just those essential for pluripotency. Once they have identified them, Agata said, "We may be able to return differentiated cells to a state of pluripotency by introducing those genes, which in turn may lead to significant breakthroughs in regenerative medicine." Agata spoke those words in a 2003 interview. Little did he know that just a few years later, his Japanese colleague Shinya Yamanaka would prove him right.

As we have seen, the pioneering Columbia zoologist Edmund Beecher Wilson gave wide currency to the term "stem cell." Wilson was among the first scientists to attribute the stem cell's uniqueness to its distinct "duality of purpose." Some call it the "to be *and* not to be" cell-fate decision. That is, the cell usually wants it both ways: to copy itself exactly *and* to yield specialized progeny. That is the destiny of many types of stem cells, including those that help make sperm and eggs and help build and repair tissues.

Studies of fruit flies and human cells are combining to yield provocative clues about what actually occurs. In 2003, Yukiko Yamashita and her Stanford University associates showed that, with the help of a protein called APC, one of the daughter cells in the germ line stem cells of fruit flies is "launched" into the destiny of sperm making while the other daughter cell behaves just like its mother cell. What tells one daughter cell to be a stem cell and the other not to be?

Building on the work of Yamashita's group with APC, Crista Brawley and Erika Matunis, then of the Johns Hopkins School of Medicine, found that a signaling protein called Jak-STAT maintains self-renewal when that signal is turned on and off in germ line stem cells. "With a few exceptions, it is thought that once cells start down the path toward specialization, they can't go back," Matunis said. "But we've clearly shown in fruit flies that lost sperm-making stem cells can be replaced, not by replication of remaining stem cells, but by reversal of more specialized cells." In other words, either stem cells or their offspring "may function as stem cells, if they can respond to appropriate signals from the niche," according to Brawley and Matunis. The *J* in Jak-STAT stands for Janus, the two-faced Roman god of passageways. In this case, the Jak-STAT signal controls the passageway for sperm-making stem cells in fruit flies. The ability to study dedifferentiation and tissue regeneration in fruit flies is likely to give a boost to understanding the same mechanisms in mammals, including humans. That's because strategies such as these are typically conserved through evolution, through the creation of species by natural selection.

The story of APC—a protein that helps to set the stage for stem cell "duality of purpose," at least in fruit fly testes—began in 1991. That year, Bert Vogelstein and colleagues at Johns Hopkins University identified the gene for adenomatous polyposis coli—the APC gene—and linked mutations in the gene to polyps and colons of families with a rare inherited disease, familial adenomatous polyposis. The mutations cause a damaging or lethal defect in early development and play a role in initiating a malignant tumor of the colon. A separate but related research track using mouse embryonic stem cells is showing that the APC protein needs to partner with another protein, beta-catenin, to strike the right balance of cell growth and cell death in stem cells in the intestine. A team of Dutch, Japanese, and American scientists reported that a defective APC/beta-catenin signaling pathway could lie at the root of tumor formation "in the colon and other self-renewing tissues."

So it could be that changes in the levels of expression of the APC gene or the beta-catenin gene may interfere with stem cell specialization and speed up the reproduction of nonspecialized cells in the embryo. Cancer cells are primitive cells that continually mutate. They pick up a genetic growth advantage over healthy mature cells and they do not die like normal cells through programmed cell death. Over time, cancer cells overwhelm healthy cells by their sheer numbers.

It is hard to imagine how a gene in fruit flies can be altered in humans and play a role in colon cancer. In the early vertebrate development, a disruption in the ability of APC or beta-catenin genes to produce the right proteins, and in the right amounts, in effect throws a bomb into the delicate signaling network that maintains the homeostasis, or harmony, of differentiating tissues. The stem cells in these primitive organisms don't know whether or not, or when to self-renew or differentiate; they appear to make their decisions in a haphazard manner. Their indecision or poor decision making leads to a catastrophe for what are called "downstream genes," those genes affected by the APC abnormality that are critical for the proper development of the body plan. The Wnt genes and their corresponding proteins are key among these downstream genes. Wnt proteins are a family of signaling molecules that regulate cell-to-cell interactions during early growth of the embryo. They are indispensable for the development of many organisms. Wnt-1, the first gene discovered in the family, is associated with the development of breast cancer in mice. It was discovered in 1982 by Nobel laureate Harold Varmus, later director of the National Institutes of Health and the National Cancer Institute, and his colleague, Roel Nusse. The Wnt signaling pathways have been favored through evolution because they are integral to the well-being, healing capacity, and reproductive potential of fruit flies, mice, and humans, to name a few. By intentionally mutating or changing certain Wnt genes in the laboratory, scientists can re-create firsthand the flaws that occur when the organism experiences a similar mutation in nature and the consequences that result. Wnt signaling showed its importance in tissue architecture when a Cincinnati research team used Wnt along with another protein called fibroblast growth factor to create intestinal tissue from pluripotent stem cells. Laboratory-created intestinal "organoids" can help scientists understand the molecular basis for intestinal diseases like adenomatous polyposis coli and eventually generate intestinal tissue for transplantation-based therapy for patients suffering from Crohn disease, inflammatory bowel disease, and colitis. Wnt went under the media klieg lights when scientists studying skin stem cells found that it is the key gene in allowing laboratory mice to grow new skin complete with hair follicles, a feat headlined as a "baldness breakthrough." Wnt actually reprogrammed the stem cells to make hair follicles, which they won't do without it.

Evolutionary and developmental biology and genetics, which constitute a hot new field in the biological sciences dubbed "evo-devo," have pegged Wnt as a sort of Rosetta Stone for understanding the development of organs and tissues in health and disease. To take one example, the APC/beta-catenin partnership is vital for the correct formation of the cardiovascular system. When the partnership is disrupted in the development of zebra fish by a genetic mutation in APC, the zebra fish ends up with a deformed heart. The heart possesses the correct tissues, but the tissues are not positioned correctly, and the heart valves are defective. Scientists studying the problem have found that excessive Wnt/beta-catenin signaling in the zebra fish heart, a result of the APC gene mutation, is the culprit.

In 2003, research teams led by Tannishtha Reya of Duke University and Roel Nusse and Irving Weissman of Stanford

University sorted out the Wnt signaling pathway for blood-forming stem cells. Blood-forming stem cells can repopulate all blood cell lineages, including cells critical for proper functioning of the immune system. But how do they do it? What tells them to renew themselves to keep the system in harmonious balance, in homeostasis? How do they replicate themselves in a controllable manner so they don't cause leukemia? How do they create the optimal number of cells for the bloodstream to carry oxygen and fight infection? How do they stop bleeding without clogging our blood and causing strokes and heart attacks?

What tells blood-forming stem cells to both constantly renew themselves and produce just the right number of daughter cells committed to keeping specialized blood cells in balance is a simple question with a complicated answer. Reya, Nusse, and Weissman and their fellow researchers used components of the Wnt/beta-catenin signaling pathway to demonstrate that, without the appropriate Wnt/beta-catenin signals, the blood-forming stem cells are unable to respond to other growth factors that are important for their self-renewal. Now that researchers have found a way to coax stem cells to divide yet not specialize, Wnt/beta-catenin signaling can be harnessed to expand a patient's own blood-forming stem cell population, if healthy, or those of a donor. Doing so would increase the likelihood of a successful transplant.

Defects in zebra fish heart valves, mouse embryonic stem cells, and homeostasis in the human gut all were found to point to the same source: a mutated APC gene. A single gene, but one with a broad biological reach. A gene that spans evolutionary time, that furnishes forensic evidence linking flies, fish, and mammals. A gene that turned up sometime after Bert Vogelstein at Johns Hopkins University began looking at the genetics of families with a high risk for colon cancer during the 1980s.

CELLULAR AGING AND IMMORTALITY

Stem cell genes are "mission control" for the developmental program of an embryo. That's why scientists started to investigate genes in human embryonic stem cells soon after they were first isolated. Genes are made up of DNA molecules, and these molecules "are the natural language of the cell," in the words of developmental biologist Lewis Wolpert. It is in the study of molecular interactions that affect the behavior of cells and embryos that the secrets of development can be discerned.

Human embryonic stem cells were first isolated in the laboratory of James Thomson and his colleagues at the University of Wisconsin–Madison in 1998. That same year John Gearhart's laboratory at Johns Hopkins University reported it had isolated human embryonic stem cells from primordial germ cells, the cells that make sperm and eggs and are considered by biologists to be immortal because they link one generation with the next. But the fascinating story of immortal cell lines cultivated in a laboratory began eighty-four years earlier in the Rockefeller Institute laboratory of Nobel Prize–winning surgeon Alexis Carrel. On January 17, 1912, Carrel placed a slice of heart muscle from a chick embryo in a culture medium. The cells lived until 1946, two years after Carrel himself died of a heart attack. For many years, Carrel's laboratory hosted a birthday party for the historic heart cells.

Carrel's success in extending the lives of chicken heart cells and hearts from chicken embryos suggested to him that death wasn't inevitable. Aviator Charles Lindbergh was intrigued by Carrel's success in surgery and tissue culture. Like Leonardo da Vinci, Lindbergh had an abiding curiosity about the human body as well as human flight. When Lindberg contacted Carrel, the surgeon realized that he could use Lindbergh's engineering skills. The two celebrities formed a highly publicized partnership. It culminated in several scientific publications, a book entitled *The Culture of Organs,* and the "Lindbergh pump," a glass perfusion device for keeping organs alive in the laboratory. The Lindbergh pump made the cover of *Time* in 1938 in a story with the lead "Men in Black," alluding to the preferred color of Carrel's surgical gown and the black floors and bleak operating room at the Rockefeller Institute where he worked. The remarkable story of the aviator and the surgeon is told by David Friedman in his book *The Immortalists: Charles Lindbergh, Dr. Alexis Carrel, and Their Daring Quest to Live Forever.*

But science is all about reproducibility, and on that score, Carrel's cell line legacy ran into trouble. Cell biologist Leonard Hayflick of the University of California in San Francisco smelled something funny in Carrel's immortal chicken cell line. He couldn't replicate Carrel's results. In 1961, Hayflick found that human cells in culture stop dividing after about fifty doublings. "Until the early 1960s it was believed that normal cells had an unlimited capacity to replicate," he wrote in 1997. "In the early 1960s we overthrew this dogma after finding that normal cells do have a finite replicative capacity. We interpreted this phenomenon to be aging at the cellular level." Cellular aging occurs as a natural consequence of cell division, which causes the DNA at the ends of chromosomes to unravel like a shoestring that's missing the plastic sleeve on its tip. The chromosomal equivalent of the plastic sleeve is the caplike *telomere.* With each division, the telomeres become shorter until the cell dies.

Life for all of us as we age is partly a story of shortening of telomeres. We don't really know what guides this process. Yet embryonic stem cells and a distinct population of adult stem cells appear able to avoid this biochemical and biological aging process found in other cells. Imagine a time when we might be able to have drugs that could deter and even counteract some aspects of the aging process. If such a time is realized, stem cells are likely to be among the major actors bringing it about. Tissue repair is a case in point. Tissue repair is much more efficient in young than older people. Stem cells specific for, and residing in, certain tissues may be more easily called into action to produce the necessary specialized cells to heal a wound or remodel bone in young people.

A quarter century after the "Hayflick limit" of cell division was proposed, Elizabeth Blackburn and Carol Greider of the University of California at San Francisco discovered that the enzyme telomerase rebuilds telomeres by repetitively adding the building blocks of DNA. Blackburn and Greider along with colleague Jack Szostak won the Nobel Prize for their research on telomeres and telomerase in 2009. If you have plenty of telomerase in the right places, or the means to make it, you may avoid the consequences of aging, such as disease, disability, perhaps even hair loss. But could a massive infusion of telomerase throw your current biological program into reverse?

The chromosomal recapping action of telomerase is one of the factors that scientists cite when they refer to embryonic stem cell lines, including the five lines developed by James Thomson in 1998, as seemingly "immortal." When Thomson's team explored the genetic features of their embryonic stem cell lines, they reported that the lines "expressed high levels of telomerase activity" and that genetic expression of telomerase "is highly correlated with immortality in human cell lines." Thomson's research was being funded in part by Geron Corporation in Menlo Park, California, which argued that embryonic stem cells, unlike other cells, make plenty of their own telomerase. Thus, Geron stipulated, embryonic stem cells have the capacity for "self-renewal" because their chromosomes get a regular recapping. The company's expertise in the field led to the development of a drug that inhibits telomerase activity. The drug is being tested in clinical trials as an anticancer agent because cancer cells make lots of telomerase.

Human embryonic stem cell lines have not been around that long and are not that old, in terms of "immortalized" cell lines, and many of their mysteries remain to be revealed. No one knows, for instance, why factors in mouse cells keep human embryonic stem cells growing and dividing in their most primitive state. We also don't know what causes embryonic stem cells implanted in mice to cause tumors. Is telomerase involved? Is this enzyme the dark secret behind the potential immortality of a cancer cell?

So it is that telomerase activity is a two-edged sword: it is one of the genetic hallmarks of early development but is also a feature of uncontrolled growth in cancer cells. Scientists are searching for ways to employ telomere dynamics of stem cells to gain a better understanding of disease, how diseases can be prevented, and how the effects of aging can be mitigated. The ultimate goal would be to translate this knowledge into a lifelong capacity for the body to repair itself almost bionically without significant risk.

GASTRULATION AND A WOMB WITH A VIEW

The embryonic stem cell lines developed by James Thomson and his colleagues at the University of Wisconsin showed even more value in developmental biology in late 2003 by revealing the earliest moments of fetal life. In Thomson's laboratory, researchers developed a stem cell model that mimics the formation of the placenta. Thaddeus Golos and his team showed that it is possible to build a window on the workings of the developing embryo's contribution to the placenta before and when it comes into contact with the uterine wall. To build that window, they created clumps of embryonic stem cells called *embryoid bodies*. The window on these developing clumps of cells reveals in crude form the beginnings of an event described by Lewis Wolpert as ranking ahead of birth, marriage, and death in importance: *gastrulation*.

Gastrulation marks perhaps the most important developmental threshold we humans must cross to get here. Gastrulation in human embryos starts at about day 14 following fertilization, eight days or so after implantation. It begins as the exquisitely precise molding of form within a clump of cells the size of a poppy seed called the *blastocyst*, the earliest form of the embryo. The inner cell mass of the blastocyst separates into two layers: the *epiblast* and the *hypoblast*. At about the beginning of the third week of development, dividing cells pile up along a line to form a thick band in the epiblast. The band is called the *primitive streak*, a zipperlike head-to-tail, left-right axis. To some, the appearance of the primitive streak is an important boundary in discussions of the moral status of the embryo. That's because the primitive streak represents the first clearly recognizable stage of embryonic development—the time when cells that build the embryo (rather than the placenta) make their debut. Its appearance marks the real beginning of tissue and organ differentiation and development. Some countries, including the United Kingdom, have passed laws permitting research on early human embryos but prohibiting such research after fourteen days of development, which is when the primitive streak begins to show.

Gastrulation, which all vertebrates experience, is a rather complicated affair. "Movements occur simultaneously over many parts of the embryo with sheets of cells streaming past each other, contracting and expanding," wrote Lewis Wolpert in *The Triumph of the Embryo*. "It taxes the minds of determined embryologists to try and visualize what is going on." The development of the tissues and organs involves the folding of sheets of cells, not unlike the folding of paper in origami. Bubbles, bulges, and layers of cells shift and stretch and stick and maneuver, sometimes around, through, and on top of one another. By the end of their acrobatics, the cells have moored themselves to the positions assigned to them. Stem cells need moorings to work their biological magic with their genetic signaling pathways and their management of developing tissues in the early embryo. Without moorings, stem cells are adrift and encumbered in their ability to perform their natural functions.

The biological magic of stem cells perhaps can be best viewed with the emergence of the three germ layers in experimental embryoid bodies. Cells in the lower regions, the endoderm, will form structures like the lungs, liver, and the lining of the digestive tract. Cells in the middle region will form the heart, muscles, bones, and blood, and other tissues of the mesoderm. Cells on top will eventually become the nervous system, including the spinal cord and the brain, external tissues like skin and sensory structures of the eye, ear, and nose, and other tissues of the ectoderm.

The developmental program is the most profound generative program known in the animal kingdom, yielding at its completion an organism more complex than accounted for in a long string of DNA letters. That's because the program relies on environmental feedback from the things it initially creates to take the next step in its creative exercise. Take the nervous system, for instance. A few days after the neural tube is formed, the top of the tube bulges outward to form a future brain atop the narrower tube, the future spinal cord. The primitive brain cells send out feelers—fingerlike projections that make connection to the cell's neighbors. The whole neurogenetic affair is a flurry of mass migration and travel.

But what is directing the traffic? Who is in control? "How these processes could be controlled, particularly by genes, is a central problem," Lewis Wolpert wrote in 1991. Today, two decades later, we still don't know, but stem cell research continues to lift the veil of secrecy slightly higher with each passing year. The study of embryoid bodies is a key contributor to that veil-lifting renaissance. Because embryoid bodies develop in the laboratory much like the embryo in the pregnant woman, the study of embryonic stem cells and their genetic profiles can tell us something about which genes are turned on or off in the emerging embryonic development of endodermal, mesodermal, and ectodermal tissues. Genes responsible for embryoid body formation as well as for stem cell pluripotency are being hunted and tracked down. The interactions of the proteins they make also give us valuable insight into germ layer formation and why embryonic stem cells feel the need to stick together in clumps. As exciting as these prospects are, the poorly understood potential of embryoid bodies to shed light on "transdifferentiation"—the process by which stem cells of one tissue may become the cells of another tissue—may be paramount.

Transdifferentiation has been a hot-button item in developmental biology and genetics. It had been a cardinal rule in developmental biology that tissue-specific adult stem cells cannot be reprogrammed. Once a stem cell commits to a specific tissue, it can't go back in time and become a cell that arises from an entirely different germ layer. Yet the rapidly growing field of stem cell research is predicated on just that: the "plasticity" of such cells to turn back their biological clocks to when they were stem cells in the embryo. Plasticity may allow stem cells to switch over to producing cells dissimilar to those from which they originated. The fact that many laboratories have successfully reprogrammed skin cells to function like embryonic-like pluripotent stem cells suggests that cellular transdifferentiation and dedifferentiation are not only possible but comparatively easy to accomplish. Adult cells are a lot more plastic than we thought not so long ago, or at least can be made to behave that way in the laboratory.

THE FUTURE OF REGENERATION

The story of regeneration and stem cells remains open-ended. We are only at the beginning of a revolution and an unfathomable transformation in biology. That revolution, the biorenaissance, is already changing the science and practice of medicine and will have much broader implications for our world and that of our children. The biorenaissance is also about agriculture, industry, the environmental sciences, aging, energy, and national security. In the wake of 9/11, the U.S. Department of Defense began brainstorming ways to put the biorenaissance to work protecting our shores, as well as our immune systems, from future terrorist attacks—not to mention regenerating skin, muscle, tendons, and even limbs damaged or lost due to battlefield injury through its Restorative Injury Repair Program. The RIR program is a priority given the numbers of seriously injured military personnel from the Iraq and Afghanistan wars. As those wars were getting under way, Brett Giroir, then deputy director of the Defense Sciences Office of DARPA, posed an intriguing question: "This ability to regenerate limbs is present in many species, and even humans can regenerate a normal liver after removing as much as 90 percent of it during surgery. So why can't this regenerative capability be available for human limbs or the brain and spinal cord?"

Whether the human liver is the exception that proves the rule of limited regeneration or provides the wellspring of knowledge that can guide regenerative medicine is not yet known. Researchers have found that while both donors and recipients of partial liver transplants experience rapid liver regeneration following surgery, recipients do better than donors. Three months after transplant, recipients' livers regenerated to 100 percent of their ideal liver volume, compared with 78 percent among donors. Why? Do liver cells with front-row seats as the surgeon's scalpel passes by take on a new "anatomical address," like cells in a salamander's blastema? Does the site of incision become the outer perimeter for inner growth? Markus Grompe of the Oregon Health and Sciences University and Curt Civin at the Johns Hopkins Kimmel Cancer Center, among other researchers, found that stem cells taken from mouse bone marrow and transplanted into mice with liver injuries helped to restore liver function. Do stem cells in our bone marrow, which are constantly regenerating our blood cells, have the ability to help repair and regrow our liver? What role does the *c-Met* gene play? In rats, liver cells divide rapidly when liver stem cells express high levels of a gene called *c-Met*, based on research done at the Salk Institute in La Jolla, California. Animals in which the *c-Met* gene is selectively removed or "knocked out" do not repair or regenerate liver tissue. Soon afterward, scientists at Penn State College of Medicine found that *c-Met* is a marker for human liver cancer and that targeting the protein *c-Met* produces may be an effective way to treat the disease. That illustrates how a gene that appears to be important for organ regeneration can also be implicated in cancer when it escapes regulatory control. Scientists are learning that genes that promote growth of healthy cells and tissues, when mutated, can promote growth of unhealthy or malignant cells. What genes hold the same clues for tissue regeneration or cancer treatment (or both) of other organs such as the lung, pancreas, or spinal cord?

If there is a story of human regeneration research that trumps others, it may well be the discovery that the fetus gives mothers a gift of regenerative stemlike cells long before it becomes a baby. Researchers led by Diana Bianchi at Tufts–New England Medical Center found that fetal cells in the mother's uterus actually have some characteristics of stem cells. These fetal cells appear to circulate throughout the mother's body. They can be found in the largest numbers in the mother's organs and tissues that are diseased or injured, where they may serve the purpose of helping to heal. What makes the baby's stem cells go into the bloodstream of its mother? Might it be a way for the baby to aid the mother if she were injured or had a wound, thus improving the odds that both baby and mother will survive? A companion study by Hugh Taylor of the Yale University School of Medicine showed that stem cells from bone marrow can help regenerate endometrial tissue, some of the tissue that makes up the uterus. The uterus can now be thought of as "a bidirectional conduit for stem cell transfer," wrote Mary Lake Polan and Mylene W. M. Yao of Stanford University School of Medicine. With better understanding of how these cells function, the time may come soon "when the prenatal child heals the mother and perhaps in the far distant future becomes the ultimate health insurance for the whole family."

Thus it may be that the uterus, first accurately represented by Leonardo da Vinci in his anatomical drawings five hundred years ago, is a dynamic gateway to tissue and organ regeneration and repair, trafficking in stem cells as well as nurturing the next generation of a species. Yet it is the organs that flank the uterus, the ovaries, where the real biological mysteries reside. That's because it is the egg that is the lead player, and it is the bearer of eggs, the women of the world, who are the focus of new reproductive technologies and stem cell research. With its ability to reprogram the human genome and its potential to produce embryonic stem cells tailor-made for treating patients without the risks of rejection, the egg reigns supreme. Does this fact portend a future in which women in developing countries are exploited for their eggs the same way the natural resources of these countries have been exploited for individual profit? Harvard stem cell scientist Kevin Eggan reported in 2007 that his team found a new way of generating stem cells by manipulating fertilized eggs in mice. They removed the DNA from a single-cell fertilized egg and added DNA from a mature body cell, a variation of the nuclear transfer technique, which typically uses unfertilized eggs. The modified cell began growing into an embryo from which stem cells were derived. The experiment demonstrated the versatility and reprogramming power of the egg. Progress in reprogramming adult cells to behave like the pluripotent stem cells found in the early embryo, without the need for eggs to do the reprogramming, has put somatic cell nuclear transfer (research cloning) on the back burner, at least for the time being.

It took three centuries following Leonardo's pioneering studies in embryology for the central principle of embryological development to be stated accurately. That was done in 1828 by a German-speaking Estonian named Karl Ernst von Baer, a year after he discovered the human egg cell. It was called "von Baer's law" and it held, in so many words, that the development of the organism is the history of growing individuality and progressive uniqueness. As species become more complex, they give up flexibility.

The late Harvard University paleontologist Stephen Jay Gould articulated our dilemma in "What Only the Embryo Knows," published in *The New York Times* just weeks after President George W. Bush announced on August 9, 2001, that he would restrict federal funding for embryonic stem cell research to existing stem cell lines. Gould described the amazing ability of the *planarian* flatworm to regenerate a head and tail when cut in two. "This capacity for regeneration—the ability of cells at a wound site to 'dedifferentiate,' or return to a state of early embryonic flexibility—becomes progressively lost in animals that evolve greater adult complexity by von Baer's universal process of 'locking in,' with increasing specialization of parts," Gould wrote. "We have, in short, traded regenerative capacity for the undeniable evolutionary advantages of maximal complexity." If we want to retrieve the lost knowledge of dedifferentiation, he argued, we have no choice but to study, for the moment at least, the secret of stemness. And that is something only the embryo knows.

No activity in modern biology has raised the decibel level of public discourse more than the use of embryos in stem cell research and cloning. And the decibel level will not diminish until substitutes like reprogrammed adult cells are proven to work without complication. Many people, especially Americans, feel passionately about these subjects. They call up in us our deepest hopes and fears. They challenge our attitudes about where we should draw the line concerning permissible research, up to and including the creation of synthetic life. For all of their ability to maintain harmony within us, stem cells are also proving to be a force for discord in the world around us—a world in which the ideals of medicine and morality often clash.

Chapter Three

CHALLENGERS OF ETHICS

He who walks straight rarely falls.

—Leonardo da Vinci

Ancient Egypt, Greece, and Rome, and medieval Islam had their golden ages, but Italy's Renaissance was unmatched in its drive for new knowledge and understanding. The free spirit of the times was fueled by the rise of printing. The language of inquiry, learning, and discovery soon found its way into the language of common people and everyday experience. Printing armed the humanist reformers with a new weapon in their fierce culture war with traditionalists, and not only in the war of words. Printing introduced Europe and the New World to Italian notions of linear perspective and human form advanced by anatomical studies.

During the Renaissance, the human body was believed to be the residence of profound cosmic mysteries, a vessel to be divinely protected. To cut into a corpse, even for science or art, was to violate the sacred order. For Leonardo da Vinci, however, the imperative of investigation and knowledge always carried the day, religious dogma and social taboo be damned. He was determined to study human anatomy, and so, by candlelight in the crypt of a church suffused with the stench of decaying and putrid flesh, he dissected cadavers of fellow citizens of Florence. His biographer Charles Nicholl reminds us that dissection was a messy thing in Leonardo's day. It took a special genius "to make visual sense of the unfamiliar landscape of glutinous and collapsing forms" and then to draw them with such precision, beauty, and transparency, particularly under the constant threat of being discovered by the authorities. With his drawing of *The Fetus in the Womb*, science, medicine, and art were brought to bear on what Leonardo called "the great mystery": human reproduction. It was, in effect, the beginning of embryology. Joseph Needham, one of the leading historians of the field, called Leonardo the father of embryology.

At the dawn of the biorenaissance, in the heated debate about the place of biology, genetics, and stem cell research—particularly embryonic stem cell research—in contemporary life, we are once again confronted by uncertainties and taboos. Once again limits are being tested and boundaries are being crossed. Those limits and boundaries are moral as well as cultural, political as well as scientific. Which side will prevail? Who will be deemed "right" and who will be "wrong"? Where will the ethical line be drawn? Who will draw it? In the days following the highly publicized research breakthrough describing how skin cells could be reprogrammed to behave like embryonic stem cells, conservative columnist and wheelchair-bound Charles Krauthammer praised President Bush in *The Washington Post* for drawing the ethical line in the right place. "The verdict is clear: Rarely has a president—so vilified for a moral stance—been so thoroughly vindicated," he wrote, adding that "scientific reasons alone will now incline even the most willful researchers to leave the human embryo alone." Writing concurrently, liberal commentator and Parkinson's patient

Michael Kinsley had a quite different take on the discovery in *Time*. He argued that Bush's ethical line had delayed research progress and that the breakthrough would not have been possible without research on embryonic stem cells. Plus, even if all embryonic stem cell research stopped tomorrow, a "far larger mass slaughter of embryos would continue" in fertility clinics, Kinsley wrote. "There is no political effort to stop it."

Krauthammer had begun his column "Stem Cell Vindication" with a remark James Thomson made upon the discovery by his laboratory and that of Shinya Yamanaka that human skin cells could be reprogrammed to behave like embryonic stem cells. "If human embryonic stem cell research does not make you at least a little bit uncomfortable, you have not thought about it enough." Within days of Krauthammer's column, Thomson and Alan Lesher, chief executive of the American Association for the Advancement of Science, responded in *The Washington Post*. In "Taking Exception: Standing in the Way of Stem Cell Research," they observed, "Far from vindicating the current U.S. policy of withholding federal funds from many of those working to develop potentially lifesaving embryonic stem cells, recent papers in the journals *Science* and *Cell* described a breakthrough achieved despite political restrictions." The work of both research teams "depended entirely on previous embryonic stem cell research." They acknowledged the discomfort of many scientists with the notion of extracting stem cells from embryonic, yet "many of the life-changing medical advances of recent history, including heart transplantation, have provoked discomfort," they wrote. "Struggling with bioethical questions remains a critical step in any scientific advancement." The "stigma" that results from the policy "has surely discouraged some talented young Americans from pursuing stem cell research."

Princeton University molecular biologist Lee Silver said in a video interview with *The New York Times* that Thomson was "very careful about language" in writing the scientific paper from his laboratory describing the reprogrammed cells:

> This is political science, not molecular science. The political science is that we're not going to call them embryonic stem cells even though, the whole point is, they are embryonic stem cells. And we're not going to call them clones even though, in fact, they are clones. We're going to avoid those words. Those words are contentious. And by avoiding those words, everybody's going to be happy.

Hope that induced pluripotent stem cells would be fully equivalent to embryonic stem cells began to fade when substantial genetic and behavioral differences between them were revealed in a series of experiments by several laboratories. A Reuters "Faith-World" blog headline "Imperfections mar hopes for 'ethical' reprogrammed stem cells" accented the debate with the use of quotation marks around the word "ethical." In the original Reuters story, published more than three years after induced pluripotent stem cells made their debut, Harvard University's George Daley expressed his view poignantly: "It has not ever been a scientifically driven argument that iPS cells are a worthy and complete substitute for embryonic stem cells. Those arguments were always made based on political and religious opposition to embryonic stem cells."

At the heart of the debate over stem cell research is language. Language is always creating new meaning depending on what words are chosen and where, when, how, and why they are used. We are conflicted by words, words that promote the case for change and those that warn of its dire consequences. Advances in medicine and stem cell research have already challenged lexicographers to redefine once-familiar words—words like "cloning," "parenthood," even "sex" itself.

The term "stem cell research" has wider currency than it ever has in the quarter century since a scientific journal by the name *Stem Cells* was launched. At the Amsterdam School of Communications Research at the University of Amsterdam, Loet Leydesdorff examined the various meanings of the words "stem cell" in 2003. What he found was geometric growth in research literature and social science citations, as well as in stories in *The New York Times* and queries of Internet search engines, beginning in 1996. In brief, Leydesdorff's analysis confirms that stem cell research has escaped the Ivory Tower and is more and more a subject of political, social, and economic conversation in the general public. The term "stem cell" has entered mainstream public discourse largely because of the debate over the moral standing of the human embryo.

Scientific knowledge is an "esoteric" knowledge that is normally the province of a small elite, wrote Harvard's Richard Lewontin and Richard Levins in a memorial to their colleague Stephen Jay Gould, the famous evolutionary biologist and arguably one of the world's best science writers. Yet, they added, the "use and control of that knowledge by private and public powers is of great social consequence to all."

The more we learn about the human embryo and its stem cells, the more complex we find the nuances of the debate. The greater the complexity, the greater our need to speak the same language. Walt Whitman's poetry is often described as the language of democracy, a language of the people. The biorenaissance calls for such a language, a language that clarifies a complex science for public consumption. A language that adapts to a world in which the biosciences are on the move, no longer confined to research labs, fertility clinics, ethics commissions, and biology textbooks. A language that conveys a common understanding of this brave new world in which science tests the limits of knowledge as it crosses the boundaries of moral sensibility.

A language that asks not only how far *can* we go, but how far *should* we go?

METAPHORICALLY SPEAKING

In 1854 the Kansas-Nebraska Act effectively sabotaged the Missouri Compromise of 1820. The newer law established the territories of Kansas and Nebraska—the first as proslavery, the second as antislavery. Under the Missouri Compromise, slavery would have been barred in both territories. Southerners and their Congressmen opposed that idea. They feared the influence of a free Kansas on America's expansion westward. Plans for a transcontinental railroad were already under way. Such a railroad running through a free state would transport the notion of freedom throughout the West.

On August 9, 2001, President George W. Bush announced that he would restrict federal funding for embryonic stem cell research to existing stem cell lines. A few days later, a commentary appeared in the *Los Angeles Times* under the headline, "Bush's Stem-Cell Ruling: A Missouri Compromise." The commentary was written by Eric Cohen, editor of the conservative ethics and technology journal *New Atlantis,* a resident scholar at the Ethics and Public Policy Center, and a consultant to the President's Council on Bioethics. From the first, Cohen wrote, the issue of stem cells struck a national nerve, bringing into sharp relief deep divides about right and wrong, life and death, and the meaning and source of human dignity. It introduced, in embryonic form, what may turn out to be the new bloody crossroads of U.S. politics: where giant leaps forward in medical science meet deeply entrenched differences about what makes life sacred, and where the American gospel of progress meets the biblical admonition against human pride.

Preserving the nation's shared values while reining in its deepest moral divides was the challenge. Did Bush meet it with his policy of approving a limited number of cell lines for research, where the life-or-death decision had already been made? Wrote Cohen:

> For now, Bush's decision seems to have achieved what most believed to be impossible: approval from many members of both the pro-life community and the patient-advocacy and medical research community. But this may turn out to be a Missouri Compromise—an effort to find the best possible solution, for now, with larger debates and disagreements just around the corner.

Invoking the Missouri Compromise over the extension of slavery in the context of the great stem cell debate of the early twenty-first century will seem inappropriate and unjust to some. But the analogy may not be that far-fetched. "If the embryo is a human being, we have to justify using it as an implement of technology," said David Fleming, a physician who teaches medical ethics at the University of Missouri. "If we create another human life in order to kill it and use it to heal somebody else, it is slavery. We fought a civil war to prevent this." During the contentious debate in Missouri over Amendment 2, a 2006 ballot initiative designed to protect all forms of embryonic stem cell research allowed under federal law, St. Louis Archbishop Raymond Burke, now a cardinal, wrote that "our tiniest brothers and sisters . . . will be made legally the subjects, the slaves, of those who wish to manipulate and destroy their lives for the sake of supposed scientific and technological progress." Amendment 2 was narrowly passed by Missouri voters, but the statewide debate continues.

The 2001 "Missouri Compromise" of stem cell policy is symbolic of the power of words in framing the debate. Critics of biotechnology and stem cell research complain of the "commodification of life" they say these activities promote. Yet it is the "commodification of language," a product of the economic system of global capitalism and the technology it craves, that now reigns supreme all over the world. Global capitalism, in the view of many linguists, is a party to the disappearance of entire languages and the homogenization of English itself, including linguistic distinctions that could potentially help advance ethical discussions about stem cells.

Agriculture and manufacturing are common metaphors in those discussions. To opponents, stem cells will be "harvested" from human "embryo farms" to make tissues and organs for others, much as corn is harvested, collected, and shipped to market. They view the hundreds of patents worldwide involving embryonic stem cells as selling out life to the profit motive. To proponents, stem cells are critical components for "Repairing the Engines of Life," in the words of a *BusinessWeek* cover story.

Momentum is also a frequently used metaphor. To opponents, embryonic stem cell research will lead to an ethical "slippery slope" at the bottom of which is human eugenics, the selective breeding of people that became a powerful movement in Nazi Germany to cleanse society of "bad" genes. To proponents in the United States, the world leader in biomedical research risks "slipping behind" the competition in Europe and Asia unless it moves quickly to seize the science for its own advantage. To opponents, the initial report of the South Korean cloning coup in 2004 "sounds hauntingly like the 'decanting room' in Aldous Huxley's *Brave New World*—systematic, precise, unrepentant about its use of women as egg factories and human embryos as raw materials," as Eric Cohen described it in *The Weekly Standard*. To proponents, death or deep freeze is the fate of any embryo "spared by the Bush policy from the indignity of contributing to medical progress," said Michael Kinsley in *Time*.

Words are at issue, highly charged words like "cloning" and "embryo." "Please don't call it cloning!" urged Bert Vogelstein, Bruce Alberts, and Kenneth Shine in *Science*. The goal of creating stem cells for regenerative medicine "is not consistent with the term therapeutic cloning," which is inaccurate and misleading because it suggests human reproductive cloning. The general term "cloning" should be abandoned in favor of "nuclear transplantation," the authors argued, an idea that cloning opponents dismissed as a transparent trick. Scientists have gotten around public opposition to cloning by dressing it up in other terms, wrote the late conservative columnist Robert Novak.

The lexicon of science "is constantly evolving," wrote Vogelstein and his colleagues, but when scientific shorthand makes its way to the nonscientific public, "there is a potential for such meaning to be lost or misunderstood, and for the terminology to become associated with research or applications for which it is inappropriate." The legacy of the eugenics movement is perhaps the best historical case in point. "The extension of genetics to eugenics owed much of its popularity in the United States and in Germany to its use of culturally resonant metaphors," wrote Matthew Chew and Manfred Laubichler in *Science*. "Labeling people as a burden, a cancerous disease, or a foreign body *(Fremdkörper)* conveyed the 'threat' to society in terms that people could relate to in their respective historical and cultural settings." But in the field of molecular biology, they write, the use of metaphors has "helped to drive science to new insights" and made it accessible to public understanding by using words to paint a picture of how molecular processes work in the human body.

Metaphors, similes, and redefinitions may not win the debate for research advocates, said Stanford's Irving Weissman in *Scientific American,* but vague language tends to create or reinforce false associations or unjustified hopes or fears. "As soon as they start using catchphrases that don't describe what's going on, it's easier for people to say that we're cloning human beings," Weissman said. "You're always going to pay if you accept language that is incorrect."

Without a doubt, "cloning" is the recurring term that perpetuates misconceptions and stokes fear about the morality of medicine.

CLARIFYING CLONING

It could be expected that when sex moved out of the bedroom and into the petri dish, it would raise an ethical ruckus, and it did. Aldous Huxley's *Brave New World* was immediately summoned by opponents of in vitro fertilization, who said that human beings didn't need technology insinuating itself into the biologically sanctioned and sacredly anointed place of the womb. Sexual reproduction as we knew it was fine as it was. But with the birth of Louise Brown in 1978, the world changed, and so did we. Another four million such "test tube" babies have been born since then, including Molly Nash's brother, Adam. What was once regarded as an awesome feat in overcoming infertility became commonplace, making countless families happy and quieting somewhat the voices of opponents who feared the worst, namely deformed babies.

It took Dolly the sheep to regain our attention. Suddenly, creating a new life not only didn't involve sex, it didn't involve the sexual union of egg and sperm. Scientists in Scotland created Dolly in 1997 through somatic cell nuclear transfer, or SCNT, commonly referred to as cloning. To clone Dolly, scientists removed the nucleus of a cell *other* than a reproductive cell (eggs and sperm) from one sheep and transferred it into an egg whose original nucleus had been removed. Dolly had one "genetic parent," not two. She wasn't genetically unique; she was genetically identical to her "parent," thanks to the DNA from an udder cell from that "parent."

Scientists also use the SCNT technique for therapeutic cloning, and that's where things tend to get confusing. While the goal of reproductive cloning is to create a new organism, the goal of therapeutic cloning is to create stem cells to treat or cure a patient with a disease. The stem cells contain DNA that's virtually identical to the existing patient's and therefore unlikely to be rejected during transplantation.

Could the technology of reproductive cloning in animals be used to create a race of genetically identical humans? The reported cloning of human embryos by Korean researchers in early 2004 ensured that the question was not about to go away. Although the Korean study was later discredited, the initial media coverage all but guaranteed that the *Brave New World* anxieties over cloning human embryos would enter a period of prolonged debate in the United States and abroad. The scientific accomplishment claimed by Hwang Woo-suk was greeted with an outpouring of national pride in Korea. Some of the fanfare, no doubt, had its origins in the fact that no father was involved in giving birth to the alleged embryonic cell clone.

Stem cells are giving the term "raw material" a new meaning. No other raw material has in its essence the ability to alter our understanding of reproduction and conception without sexual union actually taking place. What would it mean to make babies whose parents are a cell culture? Will sex as a necessary form for procreation as we have come to know it survive this century?

There is a reasonable argument that the answer is "no," wrote general practitioner and medical journal editor Mabel Chew at the end of the last millennium. Though highly unlikely anytime soon, Chew postulated, today's technological advances "may render sexual reproduction redundant tomorrow." We have long been able to control *whether* we reproduce, she wrote. Increasingly we have the ability to control *how*, *when*, and with *what genetic material* we reproduce—an exciting opportunity or an unsettling prospect, depending on your point of view.

In the months following the successful completion of the Human Genome Project in April 2003—a journey that gave us the DNA code of human life and now permits us to make the links to the DNA of all other life on earth, and perhaps beyond—stunning breakthroughs in stem cell research cast our epochal sexual affair in a new light. Several research teams reported that they had made mouse sperm and egg germ cells out of early-stage embryonic tissue, called the blastocyst. The researchers then set about to see if they could make the world's first viable embryo from an artificially constructed egg and sperm that had been created from stem cells. Think about it: a species could then be reproduced from a single cell. That is a distant prospect, but the robustness of stem cell biology is giving a new dimension to the technology of reproduction. It is not exactly sexual, through conjugation of opposite sexes, nor exactly asexual, through somatic cell nuclear transfer, a.k.a. cloning. Perhaps it is best described as "stemsexual," the making of new life by capturing a piece of the immense creativity of the embryonic stem cell before it begins to break out and form tissues and structures. Nothing shows more clearly than these stunning results the growing human control over genetic material used in reproduction.

When Japanese stem cell scientist Shinya Yamanaka and his team announced that they had reprogrammed human skin cells to behave like the stem cells found in the early embryo, he cautioned that the technology could potentially be used to create eggs and sperm. That would allow same-sex couples to have their own genetic child. "At least in theory, it's possible," he said. That possibility moved a step closer to reality in 2009 when researchers in China used induced pluripotent stem cells to create live mice that reproduced over several generations, showing that iPS cells are just as potent as embryonic stem cells when it comes to creating viable embryos in mice.

The national conversation about sex, reproduction, and genetic engineering, while at times contentious, is barely under way. Most of the conversation is about sex selection—determining the sex offspring, with males favored by many cultures—and the concern over genetic enhancement that new reproductive technologies will offer.

Based on these stunning results, nothing shows more clearly the growing human control over genetic material used in reproduction.

DESIGNER BABIES

Already widely practiced around the world with conventional "assisted reproduction" approaches like ultrasound, sex selection challenges the "fundamental understanding of procreation and parenthood," said Princeton University law professor Robert George in testimony before President Bush's Council on Bioethics, of which he was a member. "When we select for sex we are, consciously or not, seeking to design our children according to our wants and desires." What the commission feared most was that assisted reproduction through advanced technologies like in vitro fertilization and preimplantation genetic diagnosis will move rapidly from the realm of simple sex selection to genetic enhancement. It will move from the world of parents desperate to rescue a dying child by having a healthy one, such as Lisa and Jack Nash did to save their daughter Molly, to the world of parents wanting their child to be athletically or academically or socially advantaged through embryo selection or genetic manipulation or both.

"This is the beginning of the end of sex as the way we reproduce," said biophysicist and biotech entrepreneur Gregory Stock, former director of the Program of Medicine, Technology, and Society at the University of California at Los Angeles School of Medicine and author of *Redesigning Humans: Our Inevitable Genetic Future*. "We will still have sex for pleasure but we will view our children as too important to leave to a random meeting of eggs and sperm." Stock says that people are going to want to make decisions about the genetic constitutions of their children to en-sure their health and well-being. That, in his view, is unavoidable, the only question being: Is the United States going to lead under appropriate guidelines and regulations or is it going to turn the matter over to other countries and regions eager to move ahead? Some cultural groups are already moving into this terrain. In Is-rael, among affluent, highly educated people, a couple wanting to get married may seek guidance from a rabbi. The couple may be counseled to have a series of tests done to see if they might be carriers of genes that would put their offspring at risk for one of the genetic diseases prevalent among certain Jews. The results of the tests determine whether the rabbi will support the potential marriage.

Reproduction is the most challenging realm of the new tech-nologies "because this passing of the torch of life from one genera-tion to the next is really so integral a part of how we see ourselves," Stock told the Cato Institute in a debate with Francis Fukuyama of Johns Hopkins University, author of *Our Posthuman Future,* a warning about the risks of the revolution in biology. The biggest danger is that we overreact because of the enormous social and theological issues the technologies raise. As a result, "we actually restrict the kinds of medical research that will bring about cures or treatments for various diseases that are clear and present dan-gers to large numbers of adults who are suffering from them," Stock said.

Whether Stock's optimistic or Fukuyama's more pessimis-tic version of the future is borne out, one thing is fairly certain: we aren't ready for either one. "Perhaps the most important and complex decision in the history of our species is approaching: in what ways should we improve our genetic endowment?" wrote *New York Times* columnist Nicholas Kristof. "Yet we are neither focused on this question nor adequately schooled to resolve it," he added, calling for a "post-Sputnik style revitalization of science education, especially genetics, to help us figure out if we want our descendants to belong to the same species as we do." For his conservative fellow *Times* columnist David Brooks, it's already pretty much a done deal. The fact that polls show more than 40 percent of Americans would use genetic engineering to upgrade their children mentally and physically "means that sooner or later reproduction becomes a casting call for 'Baywatch,'" Brooks wrote. "There's no way people are going to foreswear the joys of creative genetics."

Wider application of the research in the constantly expanding worlds of human in vitro fertilization and preimplantation genetic diagnosis would seem to be a long way off. But is it? A few years ago, the idea that mammalian sperm and eggs could be created from stem cells in a test tube would not have been taken seriously. Nor would the idea that sex cells could be easily summoned from raw biomaterials retrieved from early-stage embryos have been entertained. A few years later, we are reminded to be wary of making such assumptions, for scientific dogma has less staying power than it used to. Now it is even possible to imagine that assisted-reproduction technologies will one day give a same-sex couple a child who shares their genetic makeup, the way all children share the genes of their biological mother and father.

Several thousand healthy babies have been born worldwide following preimplantation genetic diagnosis, a number that is growing rapidly. To parents like the Nashes, PGD is helping to create miracle cures for children with heretofore incurable diseases. To critics of biotechnology, PGD is helping to make children "the ultimate shopping experience." Opponents of PGD and tissue typing, which matches potential donors with recipients, object because they consider it morally wrong to select the traits of offspring or to create embryos knowing that most will eventually be discarded. Some ethicists find the use of PGD for matching cord blood stem cell transplant "morally troubling." Richard Doerflinger of the Council of Catholic Bishops called such procedures "search and destroy missions."

In 2006, Robert Lanza and his colleagues at Advanced Cell Technology published an article in *Nature* claiming they had derived human embryonic stem cell lines using the PGD technique that gave Molly Nash a sibling with matching bone marrow. According to the report, they removed a single cell from eight- to ten-cell embryos (blastomeres) and grew the stem cell line from them. Lanza came under attack from Doerflinger and others when it was learned that the "biopsied embryos" were destroyed rather than returned to the freezer or allowed to develop, a fact not made clear in the *Nature* paper, certainly not to people who are not scientists.

Some scientists and ethicists questioned Lanza's approach even if could be verified and reproduced. Scientists wonder whether the stem cells created in this manner are as versatile as those created from complete embryos. Ethicists wonder whether the embryo is really unharmed. How would we know? Could the single cell that was removed produce a twin? That was once suggested by then Kansas Senator and later Kansas Governor Sam Brownback, an arch-opponent of embryonic stem cell research and a Catholic. The bottom line is that the Catholic Church is opposed to in vitro fertilization, which PGD requires, because it "entrusts the life and identity of the embryo into the power of doctors and biologists and establishes the domination of technology over the origin and destiny of the human person," according to the Church's catechism.

But is it truly more ethical to let nature always have its way with reproduction and infertility and genetic disease? The public consensus may be closer to that articulated by Norman Fost in *The Journal of the American Medical Association:* Parents who create a child to save a life "would seem to be on higher moral ground that those who procreate for the more common reasons; namely, unanticipated consequences of sexual pleasure, or selfish purposes such as enriching one's personal life, or the desire for heirs."

The story of Molly Nash and her brother Adam is a story of intervention in nature's way. To some people, it is a story of intervention in God's way, in God's plan. Lisa Nash told *The Denver Post* that they had received a letter from the Vatican excommunicating them from the Catholic Church, even though they are Jewish. Few people will argue that medicine is all about interfering with nature's way—by battling disease, restoring health, extending life, and dealing with the dark side of nature. The question of whether assisted reproduction, genetic testing, stem cell research, and research cloning interfere with God's way is bound to become even bigger. Do the Nashes have any second thoughts about the decisions they made? "If God did not want PGD to be available," Lisa Nash said, "he would not have allowed the doctors to figure out how to do this."

USING SURPLUS EMBRYOS

The decision of society to allow reproduction to begin outside a woman's uterus has expanded our range of choices and altered our destiny in ways we are just beginning to understand. It is the reality of embryos created outside the human body through in vitro fertilization that is the basis of contemporary stem cell controversies.

Surveys reflected increasing public support for stem cell research in the years following President Bush's 2001 policy announcement. Many observers saw public opinion "trending toward the rights of the afflicted," in *New York Times* columnist William Safire's words, and thereby against opponents of using human embryos for stem cell research. Eric Cohen himself had written, after looking at the results from a 2001 poll, that most Americans believe that embryonic stem cell research is both morally wrong and yet medically necessary. The paradox of religiosity and pragmatism living together in the American character was plain to see. "While most Americans may say they believe in creationism rather than evolution, on issues that directly affect their own lives, like health and protection of the quality of life, science wins," wrote Andrew Kohut of the Pew Research Center during the Terri Schiavo end-of-life struggle in 2005.

The paradox had its genesis in England, with the successful development of in vitro fertilization techniques. Without the emergence of in vitro fertilization and the routine practice of discarding surplus embryos, it's hard to see how the moral debate over using them for stem cell research would have evolved as it has. "Looking back, the significance of [in vitro fertilization] cannot be overstated," wrote Cohen and William Kristol in *The Weekly Standard.* "It is the source of embryos that are now available for research; it is the technological solution for couples seeking a biological child; and it is the crucial first step in transforming human procreation in radical new ways."

Embryos left over from IVF procedures each year are either discarded by the thousands or stored for future use in fertility clinics in the United States and overseas. Fifty-two thousand embryos in Britain were being stored in freezers in 1996. In 2003, the number of embryos in suspended animation in U.S. freezers was estimated at four hundred thousand. Researchers use these leftover embryos, which are donated by clinics with permission of the donors, to produce stem cells in the laboratory as a potential source of treatments and cures. Leftover IVF embryos supplied the biomaterial for most of the embryonic stem cell lines available to the U.S. research community through the National Institutes of Health. The logic, if not the morality, of using such embryos for something good seems unassailable to most people including potential donors. Sixty percent of the more than a thousand infertility patients with frozen embryos would be willing to donate their embryos to stem cell research, based on a survey by researchers from Duke University and Johns Hopkins University. Other people believe that using embryonic stem cells this way is morally wrong because a potential human being is destroyed in the process.

Leftover embryos in fertility clinic freezers were publicly invisible until researchers began taking an interest in them. For their part, U.S. policymakers largely ignored or avoided the issue of what became of the unused embryos created with in vitro fertilization, or who would "own" the embryos if there were a divorce, or whether it was OK to use cells from an unused embryo in medical treatment. It was the United Kingdom, where IVF science was pioneered, that set about deliberating the ethical implications of the technology with an eye toward regulating it.

Leading the deliberations was Baroness Helen Mary Warnock of the House of Lords, a moral philosopher who once taught at Oxford and Cambridge universities and ran Cambridge's Girton College. In the early 1980s, she chaired a government committee charged with laying the regulatory groundwork for her country's IVF and human embryo research practices. The world's most influential report on the ethics of embryos and fertilization, released in 1984, bears her name: "A Question of Life: The Warnock Report on Human Fertilisation and Embryology." The dispute is not "between those who hold that human embryos should never be used for research and those who hold they may always be used," Warnock wrote. "It is between those who hold they may never be used, and those who hold they may be used only subject to stringent control and regulation." The report served as the basis for the U.K.'s Human Fertilisation and Embryology Act of 1990, which allows scientists to use human embryos that have not developed beyond fourteen days for a restricted range of research. (It has been argued, of course, that the fourteen-day threshold is arbitrary, which, in abstract terms, it is. Surely it is no more arbitrary, however, than assigning the fate of stem cell lines to whether they were created before or after a day in August 2001, which reflected federal policy during the George W. Bush administration.)

The wisdom of Baroness Warnock was on tap once again in 2002, when a Select Committee of the House of Lords took on the subject of stem cell research. Committee chair Richard Harries, the Lord Bishop of Oxford, began by describing the Warnock report's position on the relative potential of embryonic versus adult stem cells and the moral status of the early embryo. Arguments on all sides of the two issues were vetted: whether to prohibit destruction of early embryos while also permitting abortion, or to use embryonic stem cells at the risk of tumor formation or chromosomal abnormalities, or to use embryonic stem cells or adult stem cells from bone marrow.

But Barnoness Warnock carried the day. She noted that the question of the embryo's moral status had not changed. Her concern was the risk that women "may be exploited into giving away whole cycles of eggs" in exchange for IVF treatment.

The eggs could then be sold to researchers who want to create embryos through research cloning, which the United Kingdom permits under circumstances of "exceptional need" but tightly controls with new Human Fertilisation and Embryology Authority regulations. "We need new regulations to cover egg donation," she insisted.

Exploitation of women for their eggs is a universal dilemma. It has prompted some women who support stem cell research, including Judy Norsigian, coauthor of *Our Bodies Ourselves,* to call for new laws designed to protect women from such exploitation. It has galvanized research opponents to call for the defense of women from "Big Biotech." Others consider those concerns exaggerated. "We allow me and women to donate kidneys or portions of their liver," wrote *Boston Globe* columnist Ellen Goodman. "People participate in all sorts of research. Is there something inherently different between allowing a woman to take the risk of childbirth and allowing her to take the much smaller risk of donating eggs that may eventually cure her child's diabetes? I don't think so." Yet in 2005 the world learned that female employees in a Korean stem cell lab had been coerced to "donate" their eggs so that Hwang Woo-suk could lay the groundwork for his claim, later proved false, to being the first to clone human embryos.

In 2003, a research group in Shanghai led by Hui zhen Sheng reported that it had devised a technique for "reprogramming" skin cells to take on the characteristics of embryonic stem cells by fusing them with rabbit eggs lacking a nucleus. Unlike transferring the nucleus from a nonreproductive cell and placing it into a human egg whose nucleus has been removed, this procedure involved placing the nucleus of an adult human cell into the egg of another mammal and stimulating the egg so that it began to divide and grow embryonic stem cells. Some call experiments like this "cowboy science" because it is easy to do in a largely unregulated environment that lures scientific adventurers but scares many investors. Should such a technique work, though, it would eliminate the need for human eggs in creating stem cell lines through therapeutic cloning. Just three years after Sheng's experiment using rabbit eggs, a team of British scientists sought regulatory approval to create hybrid embryos fusing human DNA and cow eggs. The British Academy of Medical Sciences called the research "vital" for the development of new treatments. It advised the government to approve the research as long as scientists did not grow the embryos beyond fourteen days or implant them into a womb.

The idea of creating human-animal hybrid cells is anathema to many people, including some bioethicists who support therapeutic cloning. They view the unknowns in human-animal hybrid technologies as too unsettling to contemplate. But human-animal hybrid cell research has arrived in the West as well as in the East with little of the public opposition to which Americans have become accustomed. Stanford chemistry professor emeritus and Nobel laureate Paul Berg put his finger on the schizoid nature of the American experience when he said, "We will either condemn them [the Chinese] as godless members of an evil empire, or we will say 'Hey, wait a second, we can't be left out of this race.'"

Early in the biorenaissance, in 2005, a bioethics conference was held at Rome's Regina Apostolorum Pontifical University. The "Global State of Stem Cells & Cloning in Science, Ethics Law" conference took place in the shadow of the Vatican and was being sponsored by various prolife organizations, including the Federalist Society. The society describes itself as "a group of libertarians and conservatives interested in the current state of the legal order." It was founded by law professors and students, including Supreme Court Justice Antonin Scalia when he was a law professor at the University of Chicago. The society favors a smaller federal role and shifting more power to state government, which, ironically, is exactly where stem cell research initiatives are sprouting. Eric Cohen, a protégé of Leon Kass, former chair of the President's Council on Bioethics, was among the speakers, as was William Hurlbut, a physician and consulting professor in human biology at Stanford University. Hurlbut is another Kass protégé and, like Cohen, an articulate spokesperson for research restrictions on human embryonic stem cell use.

Unlike Cohen who confines his contributions to philosophy and journalism, Hurlbut had undertaken to pursue a laboratory fix that he hoped would address the issue of embryo destruction. The idea was hatched by Hurlbut and neuroscientist Evan Snyder of the Burnham Institute in San Diego and presented to the President's Council. It amounted to this: What if a crucial gene required for the blastocyst—the early embryo—were turned off before it could make the placenta in a woman's uterine wall? Would that blastocyst be merely a disabled embryo-to-be, deserving of the same protection as any embryo-to-be, or would it be a "biological artifact" that could be a legitimate source of stem cells? If the disabled blastocyst couldn't make the placenta, it wouldn't become a fetus. No placenta, no fetus, no problem.

But at the conference in Rome, Cohen bemoaned the laboratory manipulations of eggs and sperm, particularly blastocysts that form following their union. Nature has equipped them to grow and they have a desire to grow and probably wouldn't like being sabotaged. The priests gathered there scratched their heads. Even *Slate* correspondent William Saletan was confused. "What does wanting have to do with it?" he asked rhetorically. "Either the thing can grow into a baby, or it can't. If it can, it's sacred. If it can't, all this fuzzy stuff about limited development and deformed understanding and kinds of meaning doesn't add up to a basis for withholding stem cells that might save people's lives."

Using human embryos for the cause of possibly healing and even curing human disease took its first step more than a quarter century ago with the rise of in vitro fertilization. The next steps came in fairly regular increments: the discovery of embryonic stem cells in mice in 1981, the discovery of blood-forming stem cells in mice in 1988, the discovery of human blood stem cells in 1992 and their successful use in treating blood diseases, then the dramatic isolation and cultivation of human embryonic stem cells by James Thomson of the University of Wisconsin in 1998.

In the twenty-first century, human reproduction continues to begin its journey outside the womb for infertile couples. Bioengineers are hard at work on building an artificial womb—a complete fetal support system that could enable some women medically unable to bear a fetus to have children. Such a womb would also give a fetus an alternative means of life support in the event the future mother becomes ill during pregnancy. Artificial wombs could also open new vistas into the re-creation of organ systems outside the human body, including the immune system. It's difficult to even conceive of these possibilities, but who knows what the future will bring decades from now? Artificial wombs were, after all, standard reproductive furniture in Huxley's *Brave New World.*

LIFE AND DEATH ON THE MORAL COMPASS

Issues of the comparative morality across countries, cultures, and religions that embryonic stem cell research has raised are becoming more important and widespread. Surveys and discussions of embryonic stem cell research and cloning raise questions and feelings most often rooted in beliefs, faith, or a sense of what the natural order is. It is a global debate involving believers and nonbelievers, Christians and Jews, Muslims, Buddhists, Hindus, Taoists, secular humanists, atheists, agnostics, and others. Who is to say whose moral compass is correct? Whose will prevail? Are countries with a flexible, regulated system for conducting embryonic stem cell research, such as the United Kingdom and Singapore, headed toward moral bankruptcy or simply misguided? Or are they enlightened and visionary? Are they tracking the Buddhist tradition of joining medicine with spirituality to restore health?

The religious and cultural traditions of Singapore, India, China, South Korea, and Japan are distinctly different from those of the West. The wrangling over the moral status of the early human embryo is not evident in their legislative assemblies, on their airwaves, or in their public forums. Various Asian people do not view the issue in morally absolute terms the way ethically charged issues are usually framed in the West: ethical or unethical, good or bad, black or white. Viewed from the East, spiritual consciousness has more in common with a self-renewing mountain stream than a philosophical inquiry or religious tract.

Buddhism, the everlasting quest for enlightenment, is the dominant religion of the Far East, and the Far East is where investment in stem cell research and innovation is surging. Most practicing Buddhists do not have moral misgivings about research on early human embryos. As Jens Schlieter noted in *Science and Theology News,* "There is a very positive attitude toward changing nature's course if it enhances the welfare of all living beings, and more so if it allows medical advancements." Stem cells offer the possibility of self-generated replacement tissues, which is especially important for practitioners of Buddhism. That's because tissue and organ transplants go against the Buddhist belief that one's body belongs to one's ancestral chain and should not be the source of parts to be put into other people, no matter how badly those people need them. Even Islam, with its approximately 1.5 billion adherents worldwide, views cloning for research as generally acceptable.

The question of when life begins and under what circumstances is a critical one in the debate over morality. The Confucian tradition regards the defining moment of life as birth rather than conception. "Throughout Roman Catholic Europe and in much of Christian America, religious authorities teach that a fertilized egg is already a person," wrote Robert Paarlberg of Wellesley College in *Foreign Policy*. The Presbyterian Church, like many mainline Protestant churches, approves the responsible use of fetal tissue and embryonic tissue for "vital research." The Roman Catholic Church, like many evangelical Protestant churches, equates such research with the destruction of human life.

The answer to the question of when life begins has no consensus, even in the scientific community. As Scott Gilbert observes in *DevBio*, a companion to his leading textbook *Developmental Biology*, a geneticist may argue that new life begins when the chromosomes of the sperm join with those of the egg in the egg's nucleus. A developmental biologist may argue that life begins at gastrulation, the first recognizable stage of embryonic development after the embryo is implanted into the uterus of the future mother. A neurologist may put special emphasis on the appearance of brain waves in the fetus, which normally occurs toward the end of the second trimester of pregnancy. After all, it is the absence of brain waves that marks the legal definition of death in many jurisdictions. If you accept the idea that "pulling the plug" on life support mechanisms is justifiable for a person who is brain-dead, what does that imply for a fetus whose brain has yet to generate the kind of electrical activity that can be detected by an electroencephalogram (EEG)?

Some scientists and conservative politicians think that "dead embryos"—embryos made in fertility clinics that fail to grow and cannot be used for reproduction—could become an important source for stem cells. The U.S. Senate overwhelming passed a bill supporting research on dead embryos in 2007. President Bush issued an executive order June 20 of that year supporting the research the same day he vetoed a Senate bill designed to ease restrictions on federal funding for research on unused embryos from fertility clinics, many of which will be destroyed anyway. But is a "dead embryo" clinically dead? Not necessarily, says Catholic neuroscientist and ethicist Tadeusz Pacholczyk. "In our haste to obtain what we want, we may be killing an embryo," says Pacholczyk, who is also a priest. "Who is the god that says the embryo is dead?" asked stem cell research pioneer John Gearhart.

Americans are just as preoccupied with the end of life as its beginning. Broadly speaking, the idea of death is resisted in America. No number of best-selling authors, spiritual guides, or biomedical ethicists counseling its acceptance is likely to alter that resistance in a fundamental way. "I have learned never to underestimate the capacity of the human mind and body to regenerate, even when the prospects seem most wretched," wrote Norman Cousins in his 1979 best seller *Anatomy of an Illness as Perceived by the Patient: Reflections on Healing and Regeneration.*

The mythology of regeneration and renewal is deeply embedded in the American character. How else can you explain the hours and dollars devoted to dieting, exercise, and other mandatory penances to achieve a "healthy lifestyle"? How else can you explain the march of drug ads flanking stories on the evening news? How else can you explain the profound paradox, in the eyes of some observers, that Americans spend as much on health care trying to stay alive at the end of life as they do all the years up to that point? It is revealing that "somewhere between 40 and 50 percent of the total lifetime medical dollars are spent during the last six months of life as an average statistic," according to physician and Bioethics Council member Benjamin S. Carson Sr. of Johns Hopkins University. End-of-life care is a major factor in the growth of health care expenditures in the United States. More than $2.5 trillion was spent on delivering health care in the United States in 2009, 17.6 percent of the total value of goods and services produced. The figure was up from $1.4 trillion in 2001. Federal forecasters predict health care spending will exceed $4 trillion by 2016. Should health care cost that much? Who will pay? Who will decide?

BIOETHICS COUNCILS AND COMMISSIONS

Modern medicine, especially research, "seems to have made death public enemy number 1," wrote Daniel Callahan, director of the international program at the Hastings Center for Bioethics and a senior fellow at Harvard Medical School. Where it will end no one knows, but in his book, *What Price Better Health? Hazards of the Research Imperative,* Callahan offers several guideposts to illuminate the path to a sensible future: promote the idea that research should focus on premature death; give the "compression of morbidity" a research status equivalent to that now given to the prolongation of life, persuade clinicians that helping a patient have a peaceful death is as important an ideal as averting death, and redefine medical progress. Medical progress should be redefined so that its "crown jewels" are seen not as the conquest of lethal disease and increased life expectancy but the prevention of illness and disability, the management of disability, reduction in life-ruining conditions like mental illness, and promoting healthy lifestyles.

Callahan is one of the most recognized names in the world in the field of bioethics and is perhaps more responsible than anyone for its rapid rise as an academic discipline in universities throughout the country. He helped found the Hastings Center for Bioethics in 1969 in Garrison, New York, as an independent, nonpartisan, and nonprofit bioethics research institute, the first center in the world to focus on ethical issues of medicine, biology, and the environment. The center "was and is an effort to grapple with a disturbing array of ethical dilemmas generated by technologies that seem value neutral in their creation, even while problem causing in their outcomes," wrote M. L. Tina Stevens in *Bioethics in America: Origins and Cultural Politics*. Such value neutrality lurked behind the development of nuclear physics, in her view, keeping ethical issues at bay until brought to the fore by Hiroshima and Nagasaki, thermonuclear weapons testing and the Cold War, and atomic energy.

If Daniel Callahan is the most recognized bioethicist in professional circles, Arthur Caplan and Leon Kass were the most publicly visible when Kass was chair of the President's Council on Bioethics for four years following George W. Bush's televised stem cell policy address August 9, 2001. Caplan, Hart professor of bioethics at the University of Pennsylvania, founding director of its Center for Bioethics, and once an officer at the Hastings Center, is an unabashed biotechnology enthusiast though he has cautioned not to believe the "hype" that embryonic stem cell research will produce effective therapies anytime soon. In *The Scientist*, he in effect dismissed Kass and Kass's colleagues William Kristol, Charles Krauthammer, and Francis Fukuyama, who broadly agree that people will become, in a sense, "posthuman" if allowed to engineer themselves. "The main flaw with this argument is that it is made by people who wear eyeglasses, use insulin, have artificial hips or heart valves, benefit from cell, tissue, or organ transplants, ride on airplanes, dye their hair, talk on phones, sit under lights and swallow vitamins," he countered. Caplan's response to the question he posed in a *Science* commentary "Is Biomedical Research Too Dangerous to Pursue?" is equally earthy:

> Self-esteem need not be a victim of progress. Those born as a result of the application of forceps, neonatal intensive care units, in vitro fertilization, or preimplantation genetic diagnosis do not appear to suffer from undue angst about having been artificially 'manufactured.' Nor should they. There is not anything obviously more dignified about being 'made' in the back seat of a car than there is having been conceived from a pipette and a sperm donor.

Leon Kass is anything but a biotechnology enthusiast. He bears, he says, the burden of a "late-onset, probably lethal, rabbinic gene which has gradually expressed itself." He is an ethicist who has questioned the incursion of "bio" into modern life at practically every juncture, beginning with the rise of recombinant DNA technology in the early 1970s and in vitro fertilization in the late 1970s. Kass is Harding Professor on the University of Chicago's Committee on Social Thought, a prestigious body that has featured such academic luminaries as T. S. Eliot, Hannah Arendt, Friederich Hayek, Allan Bloom, and Saul Bellow. Kass is also Hertog Fellow in Social Thought at the conservative American Enterprise Institute.

What is far more important than his academic credentials was Kass's position as chair of the President's Council on Bioethics formed by George W. Bush following his televised stem-cell policy address. Although he stepped down as chair in late 2005, Kass remains an influential conservative voice on bioethics. Indeed, the "future of the life sciences in America" may be influenced by a man who has defended his opposition to stem-cell research by what he calls the "wisdom of repugnance," wrote Jerome Groopman in the *New Yorker*, adding, "Whether repugnance really offers wisdom depends, of course, on what you find repugnant. The practice of autopsy, which made modern medicine possible, was for centuries widely considered repugnant."

As council chair, Kass's ride was a bit bumpy. The dismissal of Elizabeth Blackburn, the University of California at San Francisco scientist who was abruptly terminated in early 2004, was accompanied by charges that Kass was trying to pack the council with conservatives like Diana Schaub, whom President Bush picked to replace Blackburn. The issues addressed by Bush's Council on Bioethics were divisive and contentious and, in the eyes of some, intractable in the extreme. The council's mission, after all, was "to undertake fundamental inquiry to the human and moral significance of developments in biomedical and behavioral science and technology, to explore specific ethical and policy questions related to these developments." Those with the power to influence the debate and guide recommendations were under enormous political pressure to do just that. As a result, the council's bioethical boat appeared to list to starboard in the eyes of some. Blackburn's termination and that of council member William May prompted an "open letter" of protest to Bush written by Art Caplan and signed by more than a hundred scholars, physicians, and researchers. "The creation of sound public policy with respect to developments in medicine and the life sciences requires a council that has a diverse set of views and positions," Caplan wrote.

Kass made his name by being the Jeremiah of the so-called "biotechnology project," by speaking and writing and selling books about its perils. He is a skeptic about the agenda of bioethicists as well as about the relentless march of progress in biotechnology, and he is not alone. As bioethics was incubating at the Hastings Center in the 1970s, it examined the moral significance of organ harvesting, genetic research, and in vitro fertilization, wrote Andrew Ferguson of *Bloomberg News.* Then bioethics became professionalized, Ferguson said. "It hardened into a guild for certifying 'experts.' Suddenly bioethics was made the subject of graduate programs underwritten by pharmaceutical companies and for-profit research consortiums." In brief, "cheerleaders" have replaced, in Ferguson's estimation, the skeptics and moralists that launched the field and weighed scientific advances against moral traditions. The President's Council and its chair represented what Ferguson called "a kind of standing rebuke to the profession of bioethics as it's currently practiced."

Bioethics commissions in general have held the ethical line on stem cell research, a line that a growing number of Americans think should be moved for the sake of patients and their families but also for the sake of progress and global competition. Yet public policies crafted with the participation of bioethics commissions in many other countries are both more flexible in their support for embryonic stem cell research and more uniform in opposing human reproductive cloning, the making of a genetic copy of a human being.

Fundamental inquiries often struggle to find consensus because people have fundamentally different beliefs and experiences that account for their beliefs. Kass told *Christianity Today* that as bad as it might be to destroy a creature made in God's image, "it might be very much worse to be creating them after images of one's own." Yet in his role as council chair, it is the language of cloning in particular that Kass has urged be clarified and depoliticized:

> As it happens, there is a great deal of confusion about the terms used in discussing human cloning. There is honest disagreement about what names should be used, and there are also attempts to select and use terms in order to gain advantage for a particular moral or policy position. It is terribly important to try to be accurate and fair in the matter of language. Efforts to win the moral argument by Orwellian use of speech must be resisted. This is not just a matter of semantics; it is a matter of trying hard to call things by their right names; of trying to fit speech to fact as best one can.

Under Kass's four-year leadership of the council, language received its due, inquiry was largely open and forthright, and attempts were made to answer basic questions about human nature and human dignity via five council reports. They were the subject of compliment, and in some instances acclaim, by newspaper editorials and commentators across the political spectrum. The reports are "Human Cloning and Human Dignity: An Ethical Inquiry" (2002), "Beyond Therapy: Biotechnology and the Pursuit of Happiness" (2003), "Being Human: Readings from the President's Council on Bioethics" (2003), "Monitoring Stem Cell Research" (2004), and "Reproduction and Responsibility: The Regulation of New Biotechnologies" (2004). Some pundits were left in awe. "In terms of their importance to the future of our society, these issues rank up there with war and peace," wrote Alan Murray of *The Wall Street Journal* of "Beyond Therapy," a viewpoint shared by leading columnists like William Safire of *The New York Times*. "Profound . . . poses the big questions fairly and lays out the data for futuristic debate. Time to think about the brave new world we're rushing into," Safire wrote. Yet only five months later, the future had arrived and Safire, employing another well-traveled biotech metaphor, conceded that the "genetics is out of the bottle" in a column "Reagan's Next Victory," and that it might be good to set up a stem cell initiative at the National Institutes of Health named after the former president.

Although the great stem cell debate resonates in all the documents published by the President's Council, it is in "Reproduction and Responsibility" that language is massaged and parsed in such a way that the ethical impasse over stem cell research shows fault lines, if not breaks. As conservative commentators have noted, the rise of a largely unregulated fertility industry in the United States opened the door to using embryos left over from IVF treatment for research. The practices of extra embryo management and disposal, to say nothing of couple consent procedures, vary widely across the IVF industry, based on a survey of more than three hundred IVF clinics by Caplan and colleagues published after the council's report.

Although privately supported human embryonic stem cell research is legal in the United States, publicly funded research at the federal level was restricted to approved stem cell lines, during the Bush administration, of which no more than two dozen are available through the National Institutes of Health (NIH). (Ninety percent of the country's biomedical research supporting not-for-profit organizations like universities is funded through the NIH.) It had been generally accepted by the scientific community that the human embryonic stem cell lines approved in August 2001 by Bush for federally funded research could not be used in patients for reasons of safety. The approved cell lines had been grown on mouse cells to assist the growth of the human embryonic stem cells; mouse cells could transmit tumor viruses that could pose a threat to patients if the cells were used to treat them. That fear was confirmed in early 2005, when Ajit Varki and his research team at the University of California in San Diego discovered a contaminating animal molecule in one of the approved human embryonic stem cell lines. That single molecule could cause "deleterious immune reaction and/or rejection of the transplanted cells," they wrote in *Nature Medicine,* suggesting it may be safest "to start over" by deriving new human embryonic stem cell lines not exposed to animal cells.

Contamination aside, restricting research to a few stem cell lines is "like forcing us to work with Microsoft version 1.0 when the rest of the world is already working with 6.2," complained one stem cell scientist to *Newsweek.* The more important issue on regulating new biotechnologies, however, was expressed in the appendix of the Council on Bioethics' "Reproduction and

Responsibility" report: "We believe that this [report's] language provides a way for Congress to ban reproductive cloning, while agreeing to disagree on the question of cloning for biomedical research; such a solution would prevent attempts to create cloned children while allowing debate to continue about cloning for stem cell research and regenerative medicine."

In short, President George W. Bush's Council on Bioethics agreed to be "silent" on the issue of cloning for stem cell research. Though no consensus was reached about the "degree of respect" owed to early human embryos beyond "special respect," measures were adopted, "setting upper age limits on the use of embryos in research and limits on commerce in human embryos may be agreeable to all parties to the ongoing dispute over the moral status of human embryos." The council recommended that Congress prohibit the use of human embryos in research beyond a designated stage in their development, specifically between ten and fourteen days after fertilization, when the "embryo" consists of a few hundred cells. The British limit is fourteen days. The council also recommended that Congress prohibit the buying and selling of human embryos and prohibit the U.S. Patent and Trademark Office from issuing patents "on claims directed to or encompassing human embryos or fetuses at any stage of development."

Newly elected President Barack Obama decided not to renew Bush's presidential council when its term expired in September 2009. He established his own bioethics advisory body in the wake of his executive order, lifting the Bush restrictions on federal funding for human embryonic stem cells research the previous March. The advisory body was renamed the Presidential Commission for the Study of Bioethical Issues, and Amy Gutmann, president of the University of Pennsylvania, was appointed to chair it. In its makeup, the commission was notably less conservative than its predecessor. As we observe in chapter 6, its initial deliberations concerned the emerging field of synthetic biology, its potential benefits, risks, and threats. It did not, at least at its outset, debate the moral status of the human embryo or the ethics of research cloning versus reproductive cloning, which Bush's president's council wrestled with for years without reaching a consensus.

Indeed, the United States may be the last of the advanced industrial nations where human reproductive cloning is still legal under federal law. As of 2011, thirty-five countries representing nearly four billion people—more than half the earth's population—had policies in place that permit public funds to be spent for stem cell research using embryos donated by fertility clinics. At least twelve of the thirty-five countries with liberalized policies permit stem cell lines to be derived through nuclear transfer or research cloning. All thirty-five countries with liberalized policies had banned human reproductive cloning except the United States. Though the FDA has claimed jurisdiction over cloning procedures, no reproductive cloning ban exists in federal law in the United States. The fact that no ban exists for a practice almost universally condemned as the ultimate in human genetic exploitation illustrates well the moral and political tugging and pulling going on behind the stage of public demeanor in the United States.

That tugging and pulling accompanies an idea that isn't very old. "Western civilization justly boasts of having developed the idea and the machinery of Pluralism," wrote the historian and social critic Jacques Barzun in *From Dawn to Decadence*, a monumental survey of Western civilization since the Renaissance. "It accommodates in one polity contradictory religions, moral codes, and political doctrines, all equal in status. Nothing is said about their respective merit or value, let alone their being equal, which would be meaningless."

In short, ethical lines move all the time within the polity, subject to the dynamics of the polity—that is, politics.

Chapter Four

BAROMETERS OF POLITICS

Every action needs to be prompted by a motive.
—Leonardo da Vinci

Leonardo da Vinci called himself an *omo sanza lettere,* an unlettered man. But as his biographers note, he was not disparaging himself for his lack of a university education. He was stating his independence from stultifying tradition, the dogmas prevalent in his time and ours. He acted according to what he saw with his own eyes, not according to what others told him is there to see with, say, some training in Latin. To Leonardo, clarity meant "seeing the visual evidence of the world before him with an accuracy and insight that lead into the heart of things," wrote biographer Charles Nicholl. No knowledge was valid if it could not be derived from experience, principally from observation. The eye was master.

Eschewing dogma, rejecting the party line. Leonardo, in his declaration of independence, was a blueprint for the principles of liberty on which the oldest republic of the New World was founded. Free speech and other individual rights are cherished American ideals. Judge as you see fit. Vote according to your conscience. Elect those individuals who best represent your personal and political views and will govern accordingly within our demo-cratic society. Eject from office those individuals who don't.

Democratic political systems are just beginning to grapple with the challenges posed by bioscience and technology in the age of the biorenaissance. Embryonic stem cell research will not neatly cleave political points of view into the two camps we have come to know in America from a century and a half of Democratic and Republican politics. As matters of conscience and judgments about moral and ethical issues find their way into American politics, they have the potential to shatter tradi-tional constituencies and alignments, creating long-term red-state-versus-blue-state conflicts and conflicts within red and blue states. In the 2006 Congressional elections, voters of tradi-tionally red state Missouri sent one of the leading proponents of stem cell research, Democratic state auditor Claire McCa-skill, to the U.S. Senate. "And with her came the political issue of the future: biotechnology," wrote *Slate*'s William Saletan in *The Washington Post.* "So hold on to your hats. A new kind of issue has arrived. It's moral, it's economic and it's life and death. Biotechnology is here to stay, even if humanity, as we know it, isn't."

President Bush surely calculated that he was in a no-win po-sition on August 9, 2001, no matter what he decided about fed-eral funding for stem cell research. But the way out he took—approving the use of existing stem cell lines only "where the life and death decision has already been made"—was immediately challenged for being practically untenable, if not morally bank-rupt (the charge by many on the Christian right) or inhumane (the charge by many patient advocates wanting more research). When President Barack Obama eased the Bush restrictions on federal funding for embryonic stem cell research in March 2009, he charged the National Institutes of Health with formulating new guidelines for the research. The NIH sought to strike a policy middle ground by limiting federal funding to so-called spare embryos created through in vitro fertilization that were donated for research with the consent of the donors. Human embryos created specifically for research rather than reproduction or research cloning experiments would not be eligible for federal funding. The new NIH policy, when it was published in July 2009, elicited outcries from both sides of the political spectrum: It had gone too far, and it had not gone far enough. Political purity had become the order of the day.

The positions of Republicans Orrin Hatch and Bill Frist, during the presidency of George W. Bush, hammered home the problems of compromisers in such an era. An antiabortion senator from Utah, where American progress was made manifest in the nineteenth century by the golden spike linking the nation by railroad, Hatch shocked his Senate colleagues by coming out in favor of research cloning, the use of embryonic stem cells for regenerative medicine. He also was a sponsor of legislations for a national umbilical cord program, which was signed into law in late 2005. The former senate majority leader and a transplant surgeon senator from Tennessee, a champion of funding for bioengineering and regenerative medicine. Frist initially supported federal limitations on embryonic stem cell research that put the United States at a competitive disadvantage with other countries, including Great Britain. One might be tempted to ask how could Frist reconcile taking the donated organs out of a brain-dead child but not support federal funding for research on stem cells from a blastocyst that had not even developed a nervous system and brain in the first place. Such positions can bring political rewards or make one pay a political price. They can help get candidates elected or secure their political defeat. Such is the court of public opinion.

"Bioethics in the United States reflects U.S. culture and tends to be pragmatic, market-oriented and insular," wrote bioethicist George Annas and physician Sherman Elias in 2004. "Add embryo politics to this mix and, over the past few years, the result has been a bioethics that has become so narrow and self absorbed as to be virtually irrelevant to the rest of the world." Politics—the "art of the possible" in human affairs, according to Otto von Bismarck, the "Iron Chancellor" of Germany—will decide how far the United States walks forward into the uncertain world of stem cell research. "Embryo politics" will decide whether it is wise to do so, and how we will deal with those who forge ahead on their own.

PUBLIC OPINION AND CORN-PONE POLITICS

Three years after President Bush announced that he would restrict federal funding for embryonic stem cell research, Ron Reagan addressed the 2004 Democratic National Convention in Boston on the need to make embryonic stem cell research a national priority. The son of President Ronald Reagan spoke less than two months after his father's death with the consent of his mother, Nancy, who in previous months had become the nation's leading advocate of the new research field. His speech was billed as a nonpartisan attempt to raise public awareness of what the research could offer patients and their families.

With Reagan's appearance before thirty thousand delegates, reporters, technicians, and the Democratic Party's leadership, and with millions watching him on television, biomedicine found its way for the first time, in prime time, into a major political party's convention. As much as anything else, that gives credence to the idea of a biorenaissance, of advances in biology and genetics finding their way past the technicalities of science and the debates on ethics and entering the political realm of human activity—the street fighting of politics.

"I am here tonight to talk about the issue of research into what may be the greatest medical breakthroughs in our or in any lifetime—the use of embryonic stem cells, cells created using the material of our own bodies—to cure a wide range of fatal and debilitating diseases: Parkinson's disease, multiple sclerosis, diabetes, lymphoma, spinal cord injuries, and much more," Reagan began. In his ten-minute oration, he noted that "the tide of history" is with research supporters who are "motivated by a thirst for knowledge and compelled to see others in need as fellow angels on an often difficult path, deserving of our compassion. . . . We can choose between the future and the past, between reason and ignorance, between true compassion and mere ideology."

Reagan came under immediate and withering attack for his speech. John Kilner, the president of the conservative Center for Bioethics and Human Dignity, based in Chicago, observed that history is littered with misguided attempts to relieve suffering by cutting ethical corners. In his view, Reagan's misleading language covered up the fact that producing the cells he seeks requires cloning human beings and then destroying them, which is how the Christian right in general sees embryonic stem cell research. "Cloning and killing are too high an ethical price to pay," said Kilner, "particularly when there is another, safer way to develop the same cures"—that is, through adult stem cell research.

Democratic nominee John Kerry highlighted the issue in his acceptance speech two nights later, observing that Americans have always looked to the next horizon, pioneering human flight, space exploration, and the information technology revolution. "And now it's our time to ask: What if? What if we find a breakthrough to cure Parkinson's, diabetes, Alzheimer's, and HIV-AIDS? What if we have a president who believes in science, so we can unleash the wonders of discovery like stem cell research to treat illness and save millions of lives?"

The stem cell debate has a historical precedent in generating political heat, if not shedding light. The heat was generated by the Free Silver movement against a background of economic depression by the oratorical furnace of William Jennings Bryan. Bryan, the prairie populist from Nebraska and candidate for president, capitalized on his speech at the 1896 Democratic convention in Chicago by making clear that he wanted the United States to use silver to back the dollar. It was wrong to hold in reserve an amount of gold equal in value to all the paper money in circulation, he said. Moreover, he wanted silver to be valued such that it would benefit farmers by inflating crop prices, thus easing their debt burden. Not only were the lives and well-being of the struggling farmers at stake; nothing less than the future and moral purpose of a Christian nation was in jeopardy. Bryan invoked the passion of Christ to support his cause: "You shall not press down upon the brow of labor this crown of thorns, you shall not crucify mankind upon a cross of gold." Never had the Christian message been delivered so forthrightly in American politics. Voters rejected Bryan's righteous populism for the rose-colored "new industrial order" optimism of Republican William McKinley. A century later, however, another presidential candidate bared his soul when a reporter asked him to name his favorite philosopher. The candidate was Republican George W. Bush, his response was "Christ, because he changed my heart," and conservative voters elected him not once but twice.

In America, opinions are often tied to the inherent pragmatic streak in the American character. Mark Twain expounded on that down-to-earth mandate, and took the controversy over free silver to task, in "Corn-Pone Opinions," an essay about a black slave named Jerry. From atop his master's woodpile in a Missouri town on the banks of the Mississippi River, Jerry proclaims, "You tell me whar a man gits his corn pone, en I'll tell you what his 'pinions is." Wrote Twain:

> Man cannot afford views which might interfere with his bread and butter. If he would prosper, he must train with the majority; in matters of large moment, like politics and religion, he must think and feel with the bulk of his neighbors, or suffer damage in his social standing and in his business prosperities.
>
> We all do no end of feeling, and we mistake it for thinking. And out of it we get an aggregation which we consider a boon. Its name is Public Opinion. It is held in reverence. It settles everything. Some think it the Voice of God.

It is not an exaggeration or an irreverence to say that surveys of public opinion hold a scriptural status in American politics. Political elections bear it out, to the chagrin of people enlightened or burdened by civic ideals or animated by individualism or religious fervor. Public opinion surveys of embryonic stem cell research and so-called cloning are no exception. They may bring this reality into sharper focus than almost any other issue. "Interest groups, advocates, and policymakers on both sides of the debate have taken advantage of poll results to support their claims that the public backs their preferred policy outcomes, and the competing camps have staged ongoing public communication campaigns in an effort to shape public opinion," wrote Matthew Nisbet, then at Ohio State University and now at American University, who has studied the issue extensively in the larger context of the new biopolitics. Public opinion surveys help to draw media attention, and media attention can frame an issue "in dramatic terms," according to Nisbet and colleagues Dominique Brossard and Adrianne Kroepsch of Cornell University. "Only in 2001, when the issue received heavy attention from Congress and the president, did media attention peak," they wrote in the *Harvard International Journal of Press/Politics*.

The environment surrounding the issue of stem cell research in the United States remained largely unchanged until 2004. The death of President Reagan from complications of Alzheimer's disease in June of that year galvanized support for his widow Nancy's advocacy of the research, which she had expressed publicly just one month earlier. In the weeks after Reagan's death, 206 members of the House of Representatives (thirty-six Republicans) and fifty-eight members of the Senate (fourteen Republicans) appealed to President Bush to ease the restrictions he had imposed on federal funding for embryonic stem cell research. In early 2005, Representative Mike Castle, a Delaware Republican, and Representative Diana DeGette, a Colorado Democrat, introduced the Stem Cell Research Enhancement Act, a bill that would boost federal funding for stem cell research and ease the restrictive federal policy. The bill had 186 cosponsors, including twenty Republicans. Tracking experts predicted the legislation would eventually secure a majority of the House. With a majority of senators supporting liberalization of federal stem cell policy, chances seemed good that it would pass the both houses of Congress. Editorialists around the country had voiced support for the former first lady and her efforts. Newspaper columnists on the left and the right also chimed in. Ellen Goodman of the *Boston Globe* wrote, "Stem cells may not be an instant 'cure' for Alzheimer's or Parkinson's or diabetes. But as Nancy Reagan said, 'I just don't see how we can turn our backs on this. . . . We have lost so much time already and I just really can't bear to lose anymore.' This is the final one to win for the Gipper. And his widow." *Washington Post* political journalist David Broder wrote, "In this era of intense partisanship, it is rare to see congressional Republicans and Democrats join hands, even in a humanitarian cause. Credit Nancy Reagan for helping to spur this political marvel." *New York Times* columnist William Safire observed that "if public opinion, already trending toward the rights of the afflicted, can be affected by the association of the warmly remembered Reagan name with a federal impetus to stem cell research and rigorous cloning control, I say it's a good thing. If such regulatory legislation passed by Congress included a Reagan Biomedical Research Initiative at N.I.H., President Bush should feel comfortable in signing it."

But the "trending" did not mean that opponents were throwing in the towel. Many of Nancy Reagan's one-time friends and supporters abhorred the linking of her late husband's name with stem cell research. They seized on the acknowledgment by leading Alzheimer's researchers, disclosed in a *Washington Post* story by Rick Weiss, that embryonic stem cells were less likely to prove therapeutic in the disease that stole Reagan's memory and killed him than in other diseases. William P. Clark, a national security adviser and secretary of the interior under Reagan, wrote a strong counterpoint in *The New York Times*. Clark expressed no doubt that Reagan "would have urged our nation to look to adult stem cell research—which has yielded many clinical successes—and away from the destruction of developing human lives, which has yielded none. Those who would trade on Ronald Reagan's legacy should first consider his own words." Clark argued that Reagan "would also have questioned picking the people's pocket to support commercial research."

In the eyes of research opponents, the biotech financiers and entrepreneurs were earning the reputation of those indicted by William Jennings Bryan in his "Cross of Gold" speech, "the few financial magnates who, in a back room, corner the money of the world." The opponents included Michael Reagan, who called Clark "one of my dad's closest friends" and that "those arguing for embryonic stem cells have embarked on a campaign of disinformation, claiming that there are scientific reasons for believing that their research can be expected to lead to a cure for Alzheimer's disease." Such disinformation helps generate public support for "the biotech political agenda," in the view of Michael Reagan, who expressed outrage at his half-brother Ron's appearance at the Democratic convention advocating embryonic stem cell research.

The spirit of Reagan, heralded champion of free enterprise, entrepreneurship, and government deregulation, was not to do the bidding of biotechnology if opponents had their way, even if the corporeal Reagan had done the bidding of military technology defense contractors over his two terms. Indeed, "picking the people's pocket to support commercial research" is one of Reagan's legacies, embodied in the massive defense buildup to beat the Soviet Union in the Cold War. Most Americans think it was money well spent. Public funding is already at work in developing biodefense strategies in our age of bioterror. Stem cell biologists and engineers are receiving public funds from the Department of Defense to do applied research in immunology, vaccinology, microbiology, neuroscience, genetics, and other fields immersed in stem cell research. That has implications not only for human embryonic life, but also for human life on Earth. The evil empire of what is being called our "posthuman future" by biotechnology critics is no longer conveniently headquartered in a collection of palaces and office buildings called the Kremlin. The empire is a largely invisible global network. And the genetic technologies of our posthuman future are already being recruited to address what Reagan made his principal task, "the preservation of America as a free land."

That task has always been tied to the view of America as the "shining city on a hill." It is a biblical view first enunciated by Puritan minister John Winthrop in a sermon on the deck of the *Arabella* off the Massachusetts coast in 1630. It is a view often championed by Reagan including in his farewell address to the American people, and by other advocates of "American exceptionalism," a term coined by Alexis de Tocqueville on his visit to America from France in the nineteenth century. Yet one hundred years after Winthrop arrived in Massachusetts aboard the *Arabella,* another ship with the same name delivered a cargo of African slaves to the Maryland shore. And when Irish tenor Ronan Tynan sang "Amazing Grace" at Reagan's state funeral, probably few if any of the dignitaries present considered that the great redemptive hymn was written by John Newton, an eighteenth-century born-again Christian who became a highly successful slave trader. To Newton and most of his contemporaries, slavery was an obvious feature of the divinely ordained order. That it was nicely profitable was just further evidence that it was heaven-sent. Winthrop's "shining city on a hill" was a bastion of Puritanism and profit, of "Congregationalism and capitalism." Is the American informal empire of multinational corporations,

Hollywood movies, and TV evangelists "so very different from the early British Empire of monopoly trading companies and missionaries?" asks Harvard historian Niall Ferguson, a native of Scotland.

History is full of contradictions. It does not provide clear guidance about what to do when faced with weighty moral decisions arising from the onward march of technology. It does not easily attend to issues like those surrounding embryonic stem cell research or the selection of embryos following preimplantation genetic diagnosis, as occurred in efforts by Lisa and Jack Nash to save their daughter, or the genetic enhancement technologies that are coming. History has nothing to say about what will happen if American exceptionalism in science and medicine is left fallow and even exported from the "shining city on a hill" to the country that Winthrop fled or to rising Asian powers seeking to emulate America's medical and scientific prowess. What is certain is that that history is in the hands of American legislators and politicians. As of mid-2006, said Ingrid Eisenstadter in *Barron's Online,* there were thirty-one state laws "either allowing or prohibiting human embryonic stem cell research. Don't look for lots of M.D.s or Ph.D.s among the politicians sponsoring these bills." Many of these politicians, she added, "are wringing the life out of the research [stem cell research on blastocysts], even though they would not know a blastocyst if they saw one—indeed could not tell you for sure if there is any genetic material in applesauce. Or kitty litter."

CLONING AND THE UN

Several days after the funeral of Ronald Reagan, President Bush sent a message that constituted his response to the pleas of Reagan's family that he eased restrictions he imposed on federal funding for embryonic stem cell research. "Life is a creation of God, not a commodity to be exploited by man," Bush told the largest and most conservative major Protestant denomination at its annual convention via satellite from the White House. He did not mention stem cell research by name, but no one at the Southern Baptist Convention was in the dark about what he meant. Politics is all about positioning, position of contending and opposing forces. By waging the battle of position, Bush and others hoped to right the course of American thinking. His remarks, delivered in the heat of the campaign for his reelection, came on the heels of the Democratic weekly radio address by his opponent, John Kerry, in which Kerry asserted somewhat hyperbolically that as many as one hundred million Americans have illnesses that one day could be cured or treated with stem cell therapy. "The medical discoveries that come from stem cells are crucial next steps in humanity's uphill climb," Kerry said.

What was truly extraordinary about Bush and the great stem cell debate that began during his first presidential term was not that Bush sought to bring America around to his position, but that he sought to bring the *world* around to his position. It must be, as he saw it, that what is genuinely admired about America from abroad is not its military strength and not its material well-being and standard of living, its world-class companies and successful entrepreneurs, or its universities and public health system. In Bush's view, people around the world admire America for its moral rectitude, its spiritual legacy from John Winthrop.

Why else would the White House lobby the United Nations to ban all forms of cloning, including cloning for regenerative medicine, at the expense of a ban on human reproductive cloning?

Why else would it push for such a ban when the United States stands practically alone among advanced industrial nations in permitting cloning to make babies? Current U.S. policy suggests the good (a ban on reproductive cloning) is indeed an enemy of the perfect (a ban on both reproductive and research cloning). Beginning in 2001, Republican senator Sam Brownback of Kansas repeatedly introduced human cloning legislation that would have explicitly banned both reproductive cloning and somatic cell nuclear transfer experiments (research or therapeutic cloning). The proposed legislation was repeatedly stalled by senators who objected to what they viewed as the obstruction of a promising medical technology.

In December 2001, the United Nations' General Assembly passed a resolution establishing an ad hoc committee to consider "the elaboration of an international convention against the reproductive cloning of human beings." Such a convention would become legally binding in an individual country only after the country agreed to it and signed it. But conventions carry international weight that affects even nonsignatories and can erode their resistance overtime. The matter was turned over to a working group, advised by an international group of genetics and bioethics experts, until the fall of 2003, when more than sixty nations led by the United States and Costa Rica attempted to widen the plan to include therapeutic or research cloning. "Once cloned human embryos are available, it would be impossible to control completely what was done with them," the U.S. mission argued in its position paper in August 2003:

> Stockpiles of cloned human embryos could be produced, bought and sold without anyone knowing it. Once available, implantation of cloned human embryos would take place. The tightest

regulations and strict policing would not prevent or detect the birth of cloned babies. A limited or partial ban would therefore be a false ban, creating the illusion that reproductive cloning had been prohibited.

In short, to ban human reproductive cloning effectively, "*all* human cloning must be banned.*"* That argument lost by one vote in a General Assembly legal committee deliberating the issue when Iran, representing the fifty-six-member Organization of the Islamic Conference, successfully led a procedural motion to table the matter for two years, until September 2005. The vote, taken in November 2003 after two years of what *The Scientist* described as "contentious wrangling," was eighty to seventy-nine with fifteen abstentions. Among the countries in the majority were Argentina, Belgium, China, Japan, Mexico, Russia, South Korea, Sweden, and the United Kingdom, which supported a resolution offered by Belgium that would leave the decision about therapeutic cloning to individual member nations. The two-year delay was necessary "to give adequate time for all member states to study all aspects and ramifications . . . [of] a very complex and delicate question," according to Iranian legal adviser Mostafa Dolatyar.

What was truly intriguing was the consolidation of primarily Muslim nations to defeat a U.S.-led effort on a moral issue. Therapeutic cloning "is acceptable universally by all the Shia and the Sunni Muslims," said Abdulaziz Sachedina, an expert on Islamic bioethics at the University of Virginia's Department of Religious Studies. "Embryos don't have the same sanctity [that they do in the Christian faith]. They are not regarded as a person in any sense." The only references to Islam in the voluminous reports published by the President's Council on Bioethics deal with sex selection; in particular, the report titled "Beyond Therapy" notes that, in traditional Islam, "parents are expected to continue bearing children until they have at least one son." Yet Sachedina had testified before President Bill Clinton's National Bioethics Advisory Commission on the Islamic view of stem cells and cloning for research. Within this fast-growing religion, "research on stem cells made possible by biotechnical intervention in the early stages of life is regarded as an act of faith in the ultimate will of God as the Giver of all life, as long as such an intervention is undertaken with the purpose of improving human health," Sachedina told the commission in 2000.

In 2004, all 191 UN members agreed that human reproductive cloning should be banned. But they were divided on whether to allow or ban therapeutic or research cloning. The month following the effort led by the United States to ban all forms of cloning, the General Assembly overturned the decision by its legal committee, calling for a two-year delay on any discussion of a global convention against cloning.

The United Kingdom was infuriated. Therapeutic cloning is permitted by law in the United Kingdom subject to obtaining a license from the government. "I wish to make clear that the United Kingdom would never be party to any convention which aims to introduce a global ban on therapeutic cloning, neither will the U.K. participate in the drafting of such a convention nor apply it in its national law," the country's ambassador and deputy permanent representative of the U.K. Mission of the UN Adam Thomson told the General Assembly after the decision. Douglas Sylva of the Catholic Family and Human Rights Institute in the United States shot back that the U.K. is isolated in its pursuit of research cloning. "They went way out in front of the world community when they made the decision to encourage and finance such research domestically, and now they are sensing that the world community does not agree with that decision."

But just five months later, scientists from four continents gathered at the UN in an attempt to move the debate past the impasse. They called for an immediate worldwide ban on reproductive cloning and advocated support for research on the cloning of early human embryos for the development of new treatments. Bernard Siegel, a personal injury lawyer from Coral Gables, Florida, who left his practice to found the nonprofit Genetics Policy Institute and take up the cause of stem cells, led the lobbying effort. "A UN vote to ban this important scientific research would be tragic and destroy the hopes of millions suffering from Alzheimer's, Parkinson's, diabetes, cancer, spinal cord injuries, heart disease and other devastating conditions," Siegel said. Besides, given the different views about the ethics of research on early human embryos in different countries, supporters of research cloning argued, it was unlikely the United Nations would reach consensus on the issue. Policy on therapeutic cloning was best left for member nations to develop by themselves, he argued.

Within just two weeks of the UN meeting, Miodrag Stojkovic of Newcastle University submitted the first application to clone human embryos in the U.K., unleashing an outcry from opposition groups who argued that important ethical lines were, so to speak, on the line with the fertilization authority's decision. While the U.K.'s Human Fertilisation and Embryology Authority was deliberating over the first application for research cloning in the country, another independent country in its commonwealth was in the news for human reproductive cloning. Brigitte Boisselier, president of the American cloning company Clonaid, founded by the fringe Raelian religious sect, claimed that a woman from Pretoria, South Africa, was three months pregnant with a fetus that developed from a cloned embryo. Though Clonaid provided no evidence of the thirteen babies it said it had cloned, Boisselier told reporters she was following the UN debate to see if the body would declare cloning a "crime against humanity."

The denouement to the UN's wailing and gnashing of teeth over cloning took place in March 2005. The General Assembly adopted a declaration that called on governments to "prohibit all forms of human cloning inasmuch as they are incompatible with human dignity and the protection of human life." Backed by the United States and conservative Catholic countries, the non-binding resolution passed by a vote of 84 to 34, with 37 abstentions. Countries that had already invested in cloning for medical research, including Britain, Belgium, and China, immediately declared they would not honor the resolution, even though the language was ambiguous and neither the term "research cloning" nor "therapeutic cloning" was used. That didn't stop the Bush administration and its supporters from declaring victory in a forum for which it had displayed utter contempt and been handsomely rewarded at the voting booth.

Nor was the symbolic victory to be diluted by disagreements over the meaning of words and phrases. For tireless cloning opponent Wesley J. Smith of the conservative Discovery Institute, the use of the word "inasmuch" provided no escape route for research advocates. "The whole point of the declaration, as every delegate knew, was to ban 'all forms' of human cloning," Smith wrote in *The Weekly Standard.* "Moreover, if the sentence only castigated reproductive cloning, countries like the United Kingdom, the People's Republic of China, and Belgium, which bitterly opposed the declaration, would instead have been all for it."

Within weeks of the UN vote, the matter was center stage in the political drama then unfolding in Massachusetts. Governor

Mitt Romney said he would veto a bill promoting stem cell research working its way through the legislature, because it permitted the creation of embryos for research. "What I do not support is something which has not ever been done in this country yet, and that is to clone a human embryo for purposes of research," Romney said. "I believe it's ethically wrong and so do the community of nations who voted 84–34 in the United Nations to ban this practice."

UN member nations representing more than 2.6 billion people permit research or therapeutic cloning for medical research. They include the Asian giants China and India and countries as diverse as Japan, Australia, Belgium, Israel, South Korea, Sweden, and Singapore, in addition to the United Kingdom. If all people are created equal, that means that nearly half of humanity lives in countries where such research is permitted and underway, countries with varied historical experiences, high rates of economic growth, expanding scientific literacy, and targeted investment in stem cell research. Seen in that light, the UN resolution looks less like what Smith called a "breakthrough document" and more like a rearguard action serving to mollify moral concern that embryo research is getting out of hand. It didn't look as though UN blue helmets were about to take up positions at scientific laboratories and reproductive medicine centers anytime soon.

STEM CELLS AND THE LAW

Judgments about moral issues eventually find their way into politics and into law, which is always changing, even if at a snail's pace. "The life of the law has not been logic, it has been experience," wrote Supreme Court Justice Oliver Wendell Holmes Jr. in his classic commentary *The Common Law*. Law is not purely a mathematical expression of precedent and custom. Law has to take technological change into account, including the rise of industries that saw cheap labor in children. It took more than twenty years after Holmes's famous dissent in the 1918 case *Hammer v. Dagenhart* for the courts to outlaw child labor, long viewed as a perfectly legal, if not particularly moral, activity. Times changed, attitudes changed, and the laws eventually changed preventing the exploitation of children.

How will law in the future deal with human embryos created outside the body? Will these embryos have no protection in law, allowing their use with donor consent for research and the development of therapies for treating deadly diseases, including diseases afflicting children? After all, many thousands of human embryos have been legally discarded since the beginning of in vitro fertilization, their right being solely the donors' right to decide their fate. Or will these embryos be accorded full protection in law as any citizen, preventing their use for research and inviting the government into the sanctum where decisions about human reproduction are made? Or will the law be built around what some consider the embryos' "intermediate moral status" as not just cells yet not citizens with full rights? Looking ahead, the brave new world of constitutional disputes over issues like reproductive cloning and the genetic enhancement of children will raise "novel and surprising questions about how to interpret our constitutional rights to privacy, equality and free expression," wrote George Washington University law professor and Supreme Court authority Jeffrey Rosen in a *New York Times Magazine* article in 2005.

Five years later, federal policy governing human embryonic stem cell research and federal law governing human embryo research clashed in court. The stage was set for the eventual clash when Congress passed and President Bill Clinton signed the Dickey-Wicker Amendment in 1996. The amendment is the name of an appropriation bill rider that prohibits the Department of Health and Human Services, of which the NIH is part, from using appropriated funds for the creation of human embryos for research purposes or for research in which human embryos are destroyed. After James Thomson first isolated and cultivated human embryonic stem cells in 1998, the presidential administrations of Bill Clinton, George W. Bush, and Barack Obama operated under the assumption that federal funding for human embryonic stem cell research was legal as long as no federal funds were used to destroy embryos. Stem cell lines derived from human embryos that were destroyed by private parties, in theory at least, could be eligible for federal research funding. That position was implicit in stem cell legislation that passed both houses of Congress in 2006 and 2007, as we saw in the previous chapter. The Dickey-Wicker Amendment, passed before the era of human embryonic stem cell research commenced, was viewed as a potential future roadblock by only a minority of legal scholars and observers.

Then came a court challenge by two disgruntled researchers: James Sherley and Theresa Deisher. They sued in federal district court, contending that the new Obama administration federal guidelines for human embryonic stem cell research discriminated against researchers like themselves who pursue research on adult stem cells. The policy, they argued, resulted in increased competition for limited federal funding and injured their ability to compete successfully for NIH stem cell research money. Federal district court judge Royce Lamberth ruled that the plaintiffs demonstrated a strong likelihood of success in arguing that the new government guidelines violate the intent of the law about federal funding of embryo destruction. Lamberth issued an injunction against the use of federal funds for embryonic stem cell research based on the Dickey-Wicker Amendment prohibitions. The U.S. Department of Justice appealed to the U.S. Court of Appeals for the District of Columbia Circuit, which granted a temporary override of Lamberth's injunction until the case, *Sherley et al. v. Sebelius et al.* (after the Health and Human Services Secretary Kathleen Sebelius) could be heard by the court. Upon hearing the case a three judge panel of the appeals court ruled 2–1 to reverse Lamberth's decision, arguing that "the plaintiffs are unlikely to prevail because Dickey-Wicker is ambiguous and the NIH seems reasonably to have concluded that although Dickey-Wicker bars funding for the destructive act of deriving an ESC [embryonic stem cell] from an embryo, it does not prohibit funding a research project in which an ESC will be used."

U.S. scientists working in the field of human embryonic stem cell research continue to operate in uncertain territory. Human embryonic stem cell research totaled some $123 million of the approximately $1 billion allocated by the NIH for all stem cell research (human and nonhuman) in 2010. In a survey of stem cell researchers conducted before the federal appeals court decision, public policy scholar Aaron D. Levine of the Georgia Institute of Technology found that "policy uncertainty surrounding hESC [human embryonic stem cell] research in the United States has both negative scientific and economic impacts and affects scientists working with all types of stem cells." In the long run, only congressional action can resolve the uncertainty that the Dickey-Wicker Amendment casts over the future of the field in the United States. Many bright young researchers may opt for careers in genomics, neuroscience, cancer research, or other fields rather than risk having their future held hostage by shifting political winds and stop-and-go science.

In his book *Stem Cell Century: Law and Policy for a Breakthrough Technology*, written with Stephen R. Munzer, UCLA law professor Russell Korobkin sees a number of issues with which the law will have to grapple: the morality of human embryonic stem cell research and whether the government ought to fund it; the policy issues raised by therapeutic cloning and congressional attempts to ban it; the conflict between private property and the public good raised by stem cell patents, particularly patents arising from federally funded research; autonomy and informed consent rules concerning medical experimentation with human subjects; and the donation of biological materials whether they be embryos, eggs, tissues, or organs, and whether donors should be paid.

Korobkin devotes part of a chapter to the Molly Nash story, which in his view "helps to illuminate a number of legal and policy issues that appear to be on the horizon as we look forward to the stem cell century." Should the law address the interests of stem cell donors like Adam Nash, who cannot give informed consent? In the United Kingdom, Korobkin observes, dilemmas like those posed in the Nash case are referred to the Human Fertilisation and Embryology Authority (HFEA), a government agency that regulates reproductive and related technologies. In the United States, which has no such agency, parental autonomy tends to carry the day. Parents decide what is best for their children or their children-to-be.

Should stem cell treatments like the one that restored Molly Nash to health be required to receive approval in advance from the U.S. Food and Drug Administration (FDA)? In the future, companies will collect, preserve, store, treat, and manipulate the stem cells used for treatment, Korobkin writes. What liability burdens should be placed on them if something goes wrong? What if, rather than curing Fanconi anemia, the cord blood stem cells infused into the body of someone like Molly Nash cause a fatal cancer? he asks. "Under what circumstances, if any, should an intermediary that collected and treated the donor's cord blood be legally responsible for such consequences?" Looking well into the future, who will be liable if the skin cells you had reprogrammed into stem cells turn out to be ineffective or, worse, detrimental to you when you need them most? The stem cell therapy field will experience its share of failures, and the tort system will be there to serve patients with legitimate liability claims. As for striking the right regulatory balance, Korobkin favors shielding manufacturers of new stem cell treatments from legal liability "if they fully comply with all applicable FDA regulatory requirements."

One thing is certain: the world of science, including biomedical science, is racing ahead of the world of law. "Troposphere, whatever. I told you before I'm not a scientist," announced Supreme Court Justice Antonin Scalia during oral arguments in an environmental case, provoking laughter in the court. "That's why I don't want to have to deal with global warming, to tell you the truth." If the Supreme Court shies away from trying to understand the embryo and the gene, that does not bode well for balancing individual liberty and social values, protecting intellectual property and human rights, and translating research into advances in patient care. If the U.S. Supreme Court doesn't want to deal with the "Century of Biology," who will?

ARE HUMANS PATENTABLE?

In 2001, the year of Bush's first announcement on restricting stem cell research, the University of Missouri was granted a patent for a technique for cloning mammals. The university licensed the patent to Biotransplant, a leader in moving organs from one species to another—from pigs to humans, for example. Critics seized on the patent and made it an issue because the application does not exclude humans and specifically mentions human eggs. Defenders argued that the patent deals just with the process and not products. Yet the patent states that "cloned products" are covered. Critics contend that could mean embryos, fetuses, and children, though the university asserted it would not grant permission to use the patented process to clone a child. In its 1980 landmark decision on the patentability of life, *Diamond v. Chakrabarty,* the U.S. Supreme Court wrote that patents could be issued on "anything under the sun made by the hand of man." The commissioner of patents and trademarks, spotting a problem with that courtly phraseology, took the liberty of putting human beings off limits as patentable in 1987.

Two years after Bush's 2001 announcement, Rick Weiss of *The Washington Post* reported that legislative language was being crafted for an appropriations bill that would force Congress to confront the question of whether the government should issue patents on human embryos or on the medical products developed from them. Reported Weiss:

> The U.S. Patent and Trademark Office has long said it will not issue a patent on a "human being." To do so, some argue, would violate the 13th Amendment prohibiting slavery. But the patent office has not addressed the issue of exactly when a developing embryo or fetus becomes a human being and whether its policy against patenting humans reaches back before birth.

The Thirteenth Amendment to the U.S. Constitution, passed at the end of the Civil War, overturned the *Dred Scott* case, in

which the U.S. Supreme Court in 1857 held that one human being could hold a property right on another. Under the amendment, a human being *cannot* hold a property right on another. A
patent is a form of a property right.

The issue of slavery and patenting human organisms found
its way into the deliberations of President Bush's Council on
Bioethics and into law. "Every embryo for research is someone's
blood relative," political scientist and Bioethics Council appoin-
tee Diana Schaub told an audience at the conservative American Enterprise Institute in Washington, D.C. in 2004. "Today
we are forced to wonder whether mastery and slavery might assume new forms." Yet according to a 2002 Pew Research Center
survey, African Americans tend to support federal funding for
stem cell research by a margin of 48 to 37 percent. An amendment to a 2005 omnibus spending bill proposed by Florida Republican and physician David Weldon barred the U.S. Patent
and Trademark Office from issuing patents on human organisms, codifying into law the office's 1987 administrative decision. The biotechnology industry trade group BIO warned that
Weldon's amendment contained language that would jeopar-
dize investment and innovation due to lack of patent protection. With support from prolife groups as well as the U.S. Patent
and Trademark Office, Weldon's amendment passed and was
immediately challenged in court by prochoice and women's
rights groups and the State of California. Its impact on research
and innovation proved to be negligible.

The patentability of human embryonic stem cell lines, as
opposed to human organisms, has not been at issue in the
United States, but it is in Europe. The European Union's Court
of Justice delivered a preliminary opinion in 2011 that prohibited patents on human embryonic stem cell lines because they

amount to making use of human embryos as a base material. "It must therefore be agreed, if only for the sake of consistency, that inventions relating to pluripotent stem cells can be patentable only if they are not obtained to the detriment of an embryo, whether its destruction or its modification," wrote Advocate General Yves Bot in his court opinion.

Ananda Chakrabarty, a University of Illinois scientist and formerly a General Electric scientist who invented the oil-eating bacterium that was the subject of the Supreme Court test case on human patentability, still favors granting patents on life, including hybrid human life. "What is a human?" he asked in an article published in the *American Journal of Bioethics* "This is not a question of the moral dilemma to define a human but is a legal requirement as to how much material a chimpanzee must have before it is declared a part human and therefore falls under the protection of the Thirteenth Amendment. Given the potential of the cloning technology, the scenarios could be tricky and legally confusing."

AMERICAN EXCEPTIONALISM

Social conservatives in America are the driving force behind the creed of American exceptionalism. *The Economist* described America as "a nation apart." A survey published by the British weekly in 2003 showed that Americans are far more proud of their country than British, French, Italians, and Germans. Americans are also far more likely to agree with the statement "Religion plays a very important role in my life" and with the idea that the role of government is to "provide freedom to pursue goals" rather than to "guarantee no one is in need." America is exceptional in more profound ways than most people realize, according to the magazine. "It is more strongly individualistic than Europe, more patriotic, more religious and culturally more conservative." The striking irony, of course, is that many of the original immigrants to America were seeking relief from religious intolerance in Europe.

American conservatism today, joining as it does the war on taxes with the war on popular culture, with the war on external (and internal) enemies, was forged during the Cold War of the 1950s. Perhaps no single individual was more responsible for forging that consensus than the late political pundit William F. Buckley Jr., founder of the *National Review* and creator of *Firing Line*. The application of technology to biology—that is, the unleashing of individual freedom and money in a field growing with economic opportunities and ethical dilemmas alike—is testing the conservative consensus like nothing since its coming to power. Fault lines are longer and wider, radiating across the nation from the Reagan Ranch near Santa Barbara. "People who had no kind word for Nancy Reagan when she was first lady held their tongues, as a matter of courtesy, when Mr. Reagan got sick," Buckley observed on Yahoo! News. "And now these gentry have made her a champion of their current cause, which is embryonic stem cell research."

To members of Congress and the Senate supporting the research, "Mrs. Reagan is adopted as the Mother Teresa of political reform, potentially the suffragette who will bring an end to Parkinson's, Alzheimer's and diabetes," Buckley wrote, suggesting that if a congressional majority thinks the Bush policy on stem cells is wrong and shortsighted, "let it vote to change the policy." In a time of gargantuan efforts and expenditures to prolong human life, the executive arm of government should not be overextended, he wrote. "Mr. Bush would be wise to reconsider what has been thought to be the president's exclusive authority over such questions of life and death and hubris." Buckley's advice to Nancy Reagan was written before July 19, 2006, the day Bush vetoed the Stem Cell Research Enhancement Act of 2005, his first veto in his two terms in office. He vetoed virtually identical legislation a year later, his third presidential veto.

In November 2007, New York Senator Hillary Rodham Clinton named Colorado Democratic representative Diana DeGette as an adviser on stem cell research for her presidential campaign. Despite news of the scientific breakthrough in reprogramming skin cells to behave like embryonic stem cells, DeGette vowed to introduce another stem cell research bill during the second session of the 110th Congress in 2008 that would expand funding for research on donated embryos. The cell reprogramming technique is "not a substitute for embryonic stem cell research," she said, a point of view also held by James Thomson, who pioneered both techniques. "We hope Congress will override the president's veto of the Stem Cell Research Enhancement Act," Thomson wrote in a *Washington Post* commentary shortly after he and Shinya Yamanaka appeared in the media spotlight for their breakthrough in cell reprogramming. "Further delays in pursuing the clearly viable option of embryonic stem cells will result in an irretrievable loss of time, especially if the new approach fails to prove itself."

When Alexis de Tocqueville wrote his impressions of the United States in *Democracy in America,* the railroad was just making its American debut. By one account, in 1831 there were fifty-one miles of railway, in 1872 there were sixty thousand. No other invention in the nineteenth century so transformed the American landscape, moving aside traditional values and traditional peoples with its matrix of connective transportation tissue from sea to shining sea. How "American exceptionalism" deals with the political and ethical issues surrounding the emerging life sciences in an era of global communications and economic and scientific competitiveness will shape the legacy of the term de Tocqueville famously coined.

"The New World's capacity to reinvent itself—to summon up ever newer worlds from its vast expanse of space—has reinforced the odd mixture of individualism and traditionalism at the heart of American conservatism," which is the reigning American political philosophy, wrote *Economist* reporters John Micklethwait and Adrian Wooldridge in *The Right Nation: Conservative Power in America*. Today, nearly two centuries after de Tocqueville's visit, can America reinvent itself again as the value and power of knowledge gravitate from expanses of land and natural resources like oil to the "brave new world" of biology, genomics, and stem cell research? Can America recognize how the brave new world of biology cannot be neatly separated from what *Financial Times* columnist Martin Wolf called the "brave new capitalist world?" Wolf described how we are experiencing a "modern mutation of capitalism" characterized by private equity, hedge funds, surging cross-border financial traffic, and "the triumph of the global over the local."

The Puritan "city on a hill," where language, ethics, and law were marshaled to create an ideal that became a nation, can no longer keep to itself. The ideal and the passion it entails to improve one's lot in life—"to catch fish"—is envied around the world, and emulated. It is the global pursuit of the American ideal of progress and prosperity within the realm of the sciences of life that constitutes what Aldous Huxley called the "really revolutionary revolution." Competition, markets, and choice are a big and growing part of the argument over the ethics of embryonic stem cell research, cloning for research, using stem cells to create animal-human hybrid tissues for research, and testing a cell from the early embryo for genetic disease or genetic traits. Who will be the stem cell competitors of the "nation apart" around the world, and why does it matter?

Chapter Five

OBJECTS OF COMPETITION

Obstacles cannot crush me. . . . He who is fixed to a star
does not change his mind.
—Leonardo da Vinci

Though the Renaissance was a European experience, most people think of Italy when they think of that transformational period in Western civilization, particularly for its contribution to the visual arts. The Italian Renaissance was made possible by the rise of cities, trade, and business competition that put money into the pockets of merchants who could then serve as patrons for the likes of Giotto, Michelangelo, and Leonardo da Vinci. The key was competition among the artists themselves. Their *bottegas,* their studios, were experimental workshops, crucibles of innovation that teemed with talented artists, artisans, apprentices, and competition between cities.

"What is significant is the way in which the spirit of competition, always strong in Florence, seeking to beat off rivals in Genoa, Venice and elsewhere, spread from commerce to art in the thirteenth century and after," wrote Paul Johnson in *The Renaissance.* The competition sharpened as the cult of the individual artist spread, each trying to outdo the other. The goal was not just to restore ancient Greek and Roman glory through art but also to build on ancient knowledge in all spheres and surpass it. "He is a wretched pupil who does not surpass his master," Leonardo wrote. A decade after Leonardo's death in 1519, Johnson writes, Renaissance ideas and art forms were being emulated and adapted in most parts of Europe and even found their way to the New World.

As the biorenaissance takes root around the world, competition and talent once again are key. American competitiveness is the envy of the world, widely emulated as the benchmark condition for a high-growth economy. The United States has regarded competitive success somewhat as a divinely ordained birthright, yet that birthright, with its roots in biblical scripture, the colonial world of the Puritans, and nineteenth-century manifest destiny, and expressed today by the exercise of preemptive military and economic power across the globe, is under scrutiny by the bioscience community. "What we in the U.S. need to understand is that our domestic moral terrain is not readily exportable," wrote *Science* editor in chief Donald Kennedy in 2004. "U.S. politicians can't make the rules for everyone, and they don't have a special claim to the ethical high ground." Science is, after all, an international activity.

No field of human scientific inquiry displays the global diversity of ethical, cultural, and religious heritage as much as stem cell research. At the same time, in no field have governments and research universities prepared to seize the emerging field for their own competitive advantage as much as in stem cell research. Some countries are well out of the starting gate; others are trying to find the racetrack. In the United States, California is temperamentally suited and scientifically equipped to succeed in the race, which it proved with its passage in 2004 of the $3 billion Proposition

71 to fund stem cell research. Other states—Massachusetts, New Jersey, New York, Illinois, Wisconsin, Texas, Connecticut, Maryland, Missouri—are battling to align policy for what they see as a future source of health and wealth. Supreme Court justice Louis Brandeis saw the state legislatures as "laboratories of democracy" for their willingness to take innovative approaches in meeting the needs of society. When President George W. Bush placed restriction on federal funding for human embryonic stem cell research early in his administration, some states forged ahead with funding on their own, thereby demonstrating what some commentators called the "New Federalism" in life sciences policy. Public policy researchers at the Georgia Institute of Technology reported that the total of funding of U.S. states for human embryonic stem cell research in 2009 had reached a level comparable in scale to that of the National Institutes of Health, $144 million and $125 million, respectively.

For U.S. states getting into the stem cell race, the competition is not only from California. The United Kingdom, where in vitro fertilization was pioneered, has an advanced regulatory environment and government money to fund research and fund stem cell banks. Sweden is not far behind, and Singapore, Japan, South Korea, Taiwan, China, and India have also muscled up for global stem cell competition. Culture, not surprisingly, has an impact on what type of research gets done, and where. If countries in the East like Singapore, South Korea, and China, with economic clout and a growing science research enterprise, are moving ahead aggressively with embryonic stem cell research, and United States faces "policy uncertainty" due to court challenges, what does this mean? In the United States, how can any public policy on embryonic stem cells enunciated by the federal government be a unifying principle when states, regions, and universities have decided to enter research and economic development competition on their own, generating their own resources to do so? Will the United States experience "brain drain" because public and political support is lacking, even as it leads in "clusters of innovation" that attract academic, scientific, and entrepreneurial wunderkind? What will the long-term consequences be to new and advanced medical therapies?

Where will the United States wind up in the great stem cell race?

IS AMERICA LOSING ITS EDGE?

During the George W. Bush administration there was a chorus of voices in the United States calling for easing the federal restrictions it had placed on embryonic stem cell research—voices of concerned parents like Lisa and Jack Nash, of scientists like Wisconsin's James Thomson, of advocates like actor Michael J. Fox. Fox is unapologetic about his foundation's advocacy for embryonic stem cell research in studying Parkinson's, not only for its therapeutic potential for those already afflicted with the disease, but also for its potential to keep the disease from striking in the first place. "If the government were to allow funding for embryonic stem cell research, it could use its own power of oversight to apply standards to the research that people will feel comfortable with," Fox told *Business Week* magazine. "The research will happen anyway. But it will happen in other countries now, and we won't have as much input." That sentiment was echoed just before the 2006 U.S. Congressional elections by journalist and commentator David Gergen, an adviser to four U.S. presidents. There is a larger issue in which stem cell research finds itself, one that will have a compelling political impact, Gergen said. "And that is the degree to which we find ourselves increasingly, as a nation, in competition with a rising China and a rising India and other nations, which are becoming direct threats to American jobs." When the history of our era is written, observed *New York Times* foreign affairs columnist Thomas Friedman just after the 2006 elections, it will not be about the significance of the U.S. invasions of Afghanistan and Iraq, "it will be the rise of China and India."

With many of the technical successes in the field occurring outside of the United States, advocacy has expanded from disease and support groups and the medical establishment to include groups concerned about U.S. global competitiveness. There is no doubt that America faces "a serious and intensifying challenge with regard to its future competitiveness and standard of living," and what is worse, "we appear to be on a losing path." That was the dire message underlying a National Academies study, released in 2005, that found its way into Bush's State of the Union speech in 2006 and into the federal budget. The "American Competitiveness Initiative," Bush said in his speech, committed $50 billion over ten years to increase funding for research and $86 billion for research and development tax incentives. Numerous bulbs light up in Washington concerning one unassailable fact: American science and scientists need attention and support. If they don't have it, America could lose its competitive edge. Perhaps to show its mettle on the matter, Congress passed legislation with more generous funding than Bush had proposed. The president signed the America COMPETES Act into law in August 2007.

Yet moments after announcing his competitiveness initiative in his speech, Bush made clear where the United States would *not* compete: he called for legislation to outlaw "human cloning in all its forms, creating or implanting embryos for experiments, creating human-animal hybrids, and buying, selling, or patenting human embryos." Six months later, on July 19, 2006, Bush vetoed the Stem Cell Research Enhancement Act of 2005, HR 810 sponsored by Castle and DeGette that would have made more funding available for embryonic stem cell research and would have relaxed restrictions on the stem cell lines available for federal funding. A year later he vetoed similar legislation passed by Congress with significant but not veto-proof majorities. In the view of research advocates, the edge the United States might have had in the stem cell race was once again compromised. Once again, support for America's stem cell scientists was withheld. Indeed, as we saw in the previous chapter, the Dickey-Wicker Amendment hangs like the sword of Damocles over federal funding for human embryonic stem cell research in the United States.

Since 1940, American scientists have garnered more than two hundred Nobel awards, more than three times the number won by European scientists. The post–World War II investment in U.S. science, manifested through the creation of the National Institutes of Health, the National Science Foundation, and other federal funding agencies for science, set the stage for that remarkable record. Annual federal research and development funding, national defense and civilian, totaled $114.5 billion in 2009, up from $69.8 billion in 1997 and $2 billion in 1955, according to National Science Foundation data.

In the end, public interest dictates public funding, for science and everything else. The dollars spent by government on basic and clinical biomedical research must compete with requests for all other uses of federal funds. But surveys conducted by the Pew Research Center show that news of crime, health, sports, and community affairs ranks well ahead of science and technology news with the American public, even as science and technology play an ever-larger role in the nation's health, economy, and defense—indeed, its future. Moreover, issues surrounding science and technology are rarely selected in most national polls designed to elicit the nation's top public priorities. The connection of research to economic growth is not as well appreciated as its connection to public health and national defense. Perhaps most telling is the alarming fall in public support for the profession of science. A 2003 Harris Poll showed that those Americans who viewed science as a very prestigious activity has actually dropped in the past quarter century.

"The public seems increasingly intolerant of grand, technical fixes, even while it hungers for new gadgets and drugs," wrote William Broad and James Glanz in *The New York Times*. "It has also come to fear the potential consequences of unfettered science and technology in such areas as genetic engineering, germ warfare, global warming, nuclear power and the proliferation of nuclear arms." Science involving deadly pathogens, stem cells, and cloning have disturbed deeply embedded core beliefs. The "culture of life" championed by religious conservatives and the White House during the George W. Bush administration can be a formidable barrier to some areas of scientific research. In their account of the rise of global fundamentalism, including in the United States, the authors of *Strong Religion* note the irony of the current wave occurring "after the great scientific revolutions of the twentieth century—after the unlocking of nuclear power, the development of molecular biology, the replacement of Newtonian cosmology by relativity and the quantum theory." Anxiety about the future is palpable. "These are fearful times," wrote Will Hutton, a British columnist and former editor of *The Observer* newspaper. "The fall in the birth rate across the West is testimony to a growing pessimism about the future; the menaces that together seem to make the good life unattainable range from fear that science is running amok to terrorism and climate change."

More and more, science and technology constitute a molten force on the global landscape flowing to wherever they can flourish. Science does not sit still in the minds of those who teach it, want to learn it, or want to put it to work for human betterment. Science is less likely to sit still if the environment in which it is carried out is not a nurturing one. Foreign advances in basic sciences "now often rival or even exceed Americans', apparently with little public awareness of the trend or its implications for jobs, industry, national security or the vigor of the nation's intellectual and cultural life," Broad wrote in a follow-up article in *The New York Times*. Intellectual property and scientific creativity are migrating rapidly toward the center of the innovation nexus. Because the competition for scientific talent has never been greater, scientists—especially young ones—are moving on to more promising sites within an increasingly global scientific environment to pursue their careers.

Along with Singapore, England is one of the most attractive venues for American scientists wanting to work on embryonic stem cells in a regulated but unencumbered environment. England not only permits but also promotes embryonic stem cell research and research cloning. One of the research stars to depart the United States for England was Roger Pedersen, who left the University of California in San Francisco in 2001 to do stem cell research at England's Cambridge University. Cambridge has invested millions of pounds to create a world-class "nerve center" for stem cell technology, drawing talent from around the world. Pedersen wants to lay the foundation for a stem cell spin-out company modeled on Amgen, the biotechnology leader in California. "If you look at what Amgen in the U.S. has achieved, essentially by focusing on blood stem cells, then this gives some idea of the scope for creating a similar company in Cambridge," he told *Business Weekly*. To put this in perspective, consider that the global market for just one drug that makes more red blood cells from blood-forming stem cells—erythropoietin—is approximately $3 billion annually. Imagine the value of a drug that could regenerate heart muscle for people who have had heart attacks or who have heart failure for other reasons. It would be staggering.

Asian countries like Singapore also are seizing the opportunity for a "brain gain" of American scientists. Not even a generous offer from Stanford University and the prospects of Proposition 71, California's $3 billion stem cell initiative, were enough to keep the husband and wife team of Neal Copeland and Nancy Jenkins from setting up a cancer stem cell research program in Singapore in 2006. Singapore's gain was highlighted when Copeland and Jenkins were elected to the prestigious U.S. National Academy of Sciences three years later.

Whether the departure of Pedersen, Copeland, and Jenkins is representative of what regional development expert and *Creative Class* author Richard Florida called the "coming brain drain" won't be known for some time. Europe is dealing with its own brain drain of scientific talent to the United States, particularly in biotechnology. The atmosphere in Germany for basic genetics research remains thick with the remnants of Nazi-era eugenics and human experimentation. In Germany and Italy, human stem cell harvesting and cloning for any purpose are criminal acts. But countries with a favorable stem cell research climate are not necessarily safe from the lure of American science. The University of Minnesota hired Jonathan Slack from the University of Bath in England to head its Stem Cell Institute. Paul Simmons, a program director at the Australian Stem Cell Centre, moved to the United States in 2006 to become director of the Stem Cell Center at University of Texas Health Science

Center. Simmons also assumed the presidency of the International Society for Stem Cell Research. His fellow Australian Alan Trounson, founder of the Australian Stem Cell Centre, was appointed president of the California Institute for Regenerative Medicine the following year.

Carl Johan Sundberg, of Sweden's Karolinska Institute and the academic consortium Euroscience, acknowledges that Europe would benefit from a politicized atmosphere in the United States, not only by attracting additional Roger Pedersens, but also by giving young scientists from the Middle East, Asia, and Eastern Europe an increasingly attractive alternative to the United States. Some of the best non-U.S. scientists in training feel they must be educated in the United States to have the best chance of achieving the best position in academia or industry, be that in their home country or elsewhere. Yet surveys showed a broad decline in the number of foreign students applying to graduate and doctoral programs in science at American universities following 9/11. "After peaking in the mid-1990s, the number of doctoral students from China, India and Taiwan with plans to stay in the United States began to fall by the hundreds, according to the [National Science Foundation]," wrote William Broad of *The New York Times.* Heightened security concerns and new visa policies were behind the decline, which finally began to reverse in 2006 with students from China and India enrolling in U.S. graduate programs in larger numbers.

The United States, once seen as the beacon of opportunity for foreign students aspiring to become scientists and engineers, today casts a somewhat dimmer light across the world's scientific communities. That has two profound consequences, says Thomas Friedman, the Pulitzer Prize–winning foreign affairs columnist for *The New York Times*: "First, one of America's greatest assets—

its ability to skim the cream off the first-round intellectual draft choices from around the world and bring them to our shores to innovate—will be diminished, and that in turn will shrink our talent pool. And second, we could lose a whole generation of foreigners who would normally come here to study, and then would take American ideas and American relationships back home. In a decade we will feel that loss in America's standing in the world."

Sundberg contends the whole world would lose if the dynamic quality of U.S. science deteriorates. That dynamism owes more than ever to the intellectual and cultural biodiversity that nascent talent from abroad supplies. The combination of the drop in foreign students, the declining interest of American students in sciences careers, and the aging of the technical workforce, in Broad's account, prompted then American Association for the Advancement of Science president Shirley Ann Jackson to ask, "Who will do the science of this millennium?"

CLUSTERS OF INNOVATION

Where ideas in the biosciences are generated can have huge effects on regional economies and the appeal of cities and regions for highly skilled workers, their families, and cutting-edge companies. Historically, scientific progress has made its greatest leaps in geographic space, writes historian David Livingstone in *Putting Science in Its Place: Geographies of Scientific Knowledge.* Knowledge flourished in the fifteenth and sixteenth centuries on the Iberian Peninsula under the influence of Muslim and Jewish scholars, in Italy with Fracastoro's germ theory and Vesalius's study of the human body, in Poland with Copernicus's placement of the sun at the center of the universe, in Germany with Martin Luther's strategic use of print technology, and later in England with Isaac Newton's creation of the "new Logick" of mathematics and nature. Europe and America benefited from the eighteenth-century Enlightenment, which gave weight to the use of reason and organized knowledge in human affairs.

Intellectual property and technological knowledge may arise anywhere and move around the world with the speed of light, like global capital, but their chances of emerging or alighting at a nexus of intellectual ferment and expertise and profit potential are better. Urban regions are the lifeblood of the global economy. In their ventricles are the key idea generators: institutions of research, higher education, finance, and commerce. In the United States, top universities in life sciences patenting and licensing tend to be located in the heart of metropolitan regions where the biotechnology industry is growing fastest.

No one has made more hay of the growth of urban-regional economics than Michael Porter, founder of Harvard's Institute for Strategy and Competitiveness. In *The Competitive Advantage of Nations*, first published in 1990, he describes how nations, states, and regions compete and how wealth is generated in competitive "clusters" of innovation—geographically concentrated groups of companies and institutions linked by the thread of common technologies and skills. These clusters, which have flourished since the 1990s, are nowhere more in evidence than in the emerging biosciences and stem cell research. This field in particular requires the proximity of academic research faculty, technology transfer offices, management skill resources, specialized service networks, and sophisticated investor communities. Most American biotechnology companies doing stem cell research are located in California, the Boston area, and the New York–Philadelphia–Washington, D.C., corridor. Nearly all companies doing stem cell research worldwide are located in similarly dynamic urban regions. Debora Spar, president of Barnard College and author of *The Baby Business*, noted that almost all companies doing stem cell research worldwide are located in similarly dynamic urban regions and that such companies can "exert a powerful influence on how and where the stem-cell industry develops."

America's current lead in scientific infrastructure and brainpower is not a surefire claim to future industries locating on American soil. Even unregulated monopolies aren't safe from global competition, as was demonstrated when Standard Oil's early reign over an industry and market then shifted and flourished in the Middle East. But the lion's share of the new knowledge being churned out by the United States' vast research enterprise is in the life sciences. Scientists and investors know it. So do governors, mayors, and economic developers who see the life sciences as a job creation machine for their states, cities, and regions.

Wherever stem cell research is seeded and cultivated is where future industry is most likely to "grow jobs." Even though only a dozen or so private firms in the United States were active in the field of embryonic stem cell research in 2003, spending a total of just $70 million, "this still-small business is likely to expand dramatically" in the years ahead, Debora Spar predicted. Within days of the news that teams of scientists in Japan and the United States had successfully reprogrammed human skin cells to function as pluripotent stem cells, venture capitalists pulled together money and stem cell research heavyweights from Harvard and Stanford universities, the Scripps Research Institute, the Massachusetts Institute of Technology, and the University of Washington and formed Fate Therapeutics, Inc. Headquartered in San Diego, the company is developing small-molecule drugs that reprogram mature adult cells into stem cells that can repair damaged tissue and ailing organs. Robin Young, a private analyst and author of "Stem Cell

Analysis and Market Forecasts 2006–2016," predicted that sales of treatments and therapies associated with stem cells of all types in the United States would reach $8.5 billion by 2016, a conservative estimate in light of subsequent market forecasts. State lawmakers are doing their part in drawing the moral and ethical lines and taking the heat. In the process, they are marking off the boundaries of field plots for biomedical advances and regional development.

Funding restrictions on stem cell research, coupled with the rise of competitive urban-regional bioscience clusters of innovation, have unleashed a war among the states for brainpower and economic growth. In 2003, little more than two years after President Bush announced restriction of federal research funding to designated stem cell lines, state governments began seizing what they saw as a golden opportunity for their economic future. And where governments lagged behind, public and private universities stepped up to the plate, even in Texas where Bush got his start in politics. An anonymous donor gave $25 million to the University of Texas Health Science Center at Houston to create a world-class center for stem cell research. A month later, Shahla and Hushang Ansary, two Houston philanthropists originally from Iran, gave $15 million to the Weill Cornell Medical College in New York to establish a Center for Stem Cell Therapeutics. Then New York City's billionaire mayor, Michael Bloomberg, gave $100 million to his alma mater, Johns Hopkins University, for stem cell research. Government's most basic responsibility is the health and welfare of its people, Bloomberg told Hopkins' graduates in 2006, "so it has a duty to encourage appropriate scientific investigations that could possibly save the lives of millions." Donations to university stem cell research programs across the country have become routine.

New Jersey, Massachusetts, Wisconsin, and Minnesota, each boasting a critical mass of specialized research talent and first-

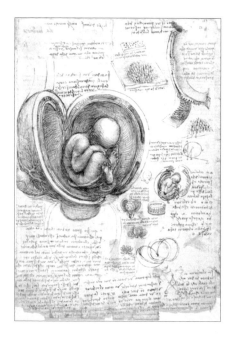

The Fetus in the Womb. Leonardo da Vinci, 1512. The drawing depicts what Leonardo called "the great mystery," a mystery in many ways more profound than the enigmatic smile on his Mona Lisa. *Photo Credit: Wikimedia Commons.*

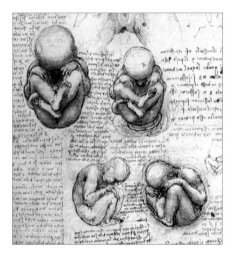

Studies of the Fetus in the Womb. Leonard da Vinci, c. 1511-13. "The navel is the gate from which our body is formed by means of the umbilical vein," Leonardo wrote. What he could not have imagined as he examined the umbilical cord attaching the fetus to the mother was that it is a treasure trove of stem cells. *Photo Credit: Wikimedia Commons.*

John Wagner, a University of Minnesota cord blood stem cell expert, holds a sleeping Adam Nash. Adam's sister Molly was born with Fanconi anemia, an often-fatal genetic disease. She received a transplant from Adam's umbilical cord blood in 2000. Adam's stem cells rebuilt Molly's bone marrow. Today, she is disease free. *Photo by Mark Engebretsen, University of Minnesota, 2000.*

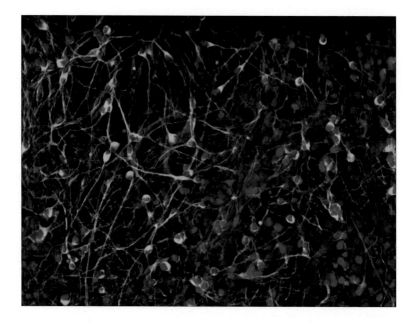

Researchers at Harvard Medical School created a brain tissue-like neural network of nerve cells (green), glial cells (red), and cell nuclei (blue) from human induced pluripotent stem cells. *Courtesy of Rakesh Karmacharya, MD, PhD, and Stephen J. Haggarty, PhD.*

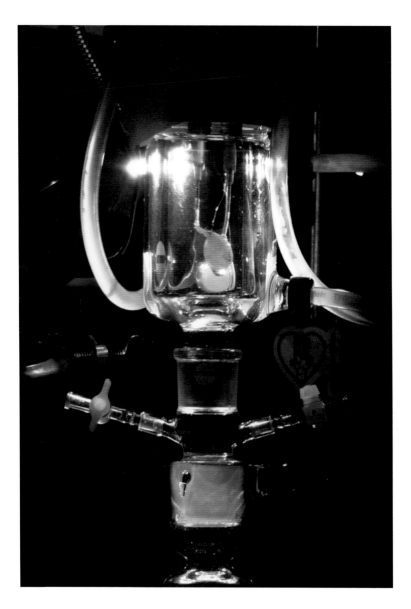

The bioengineered heart of a laboratory rat in a bioreactor at the University of Minnesota. Researchers stripped the heart of its cells, leaving behind a ghost-like protein framework. Then they seeded the framework with stem cells and progenitor cells, which rebuilt the heart's vessels and tissues. When the researchers applied an electric pulse, the regenerated heart began to beat. *Photo courtesy of Emily Jensen, University of Minnesota.*

Bioengineered lungs of a laboratory rat hang in a bioreactor. Researchers transplanted the cell-regenerated lungs into live animals where they functioned normally. *Photo courtesy of Harald Ott, MD, Massachusetts General Hospital and Harvard Medical School.*

Actor, director, producer, and screenwriter Christopher Reeve was an outspoken advocate of stem cell research and expedited clinical trials for experimental cell therapies. He died in 2004 of an infection following paralysis from a horse riding accident in 1995. *Photo by Don Flood.*

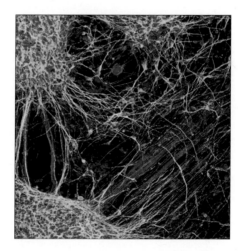

Neurons grown from skin cells of patients with schizophrenia after the cells had been reprogrammed into pluripotent stem cells. Such "disease in a dish" models could help pin down the genetic basis of many diseases. *Courtesy of Kristen Brennand, PhD, Salk Institute for Biological Studies.*

Vitruvian Man. Leonardo da Vinci, 1490. The man's navel, the exact center of his body, is the motionless point where the artistic and scientific individuality cultivated by the Renaissance had its symbolic origin—the timeless pivot of life's compass. *Photo credit: Wikimedia Commons.*

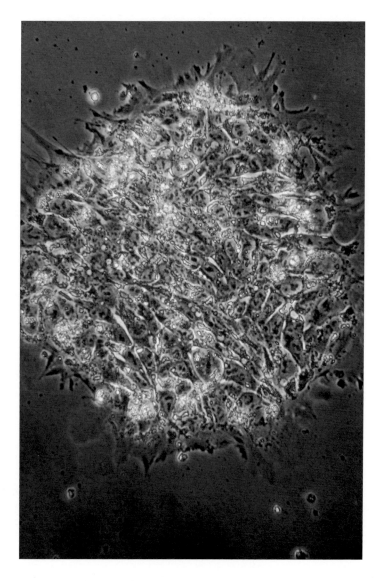

Microscopic view of a colony of the original human embryonic stem cell lines from the James Thomson lab at the University of Wisconsin–Madison. *Photo courtesy of Jeff Miller, University of Wisconsin-Madison.*

University of Wisconsin developmental biologist James Thomson established the first human embryonic stem cell lines in 1998. *Photo courtesy of Jeff Miller, University of Wisconsin-Madison.*

Japanese stem cell scientist Shinya Yamanaka of Kyoto University. Yamanaka led one of two research teams that discovered the master stem cell gene nanog, named after the Celtic Tir Nan Og or Land of Forever Young, in 2003. Then he and his colleagues pioneered the technique that creates embryonic-like pluripotent stem cells from adult cells, first with mouse skin cells in 2006, and then with human skin cells in 2007. *Photo credit: Wikimedia Commons.*

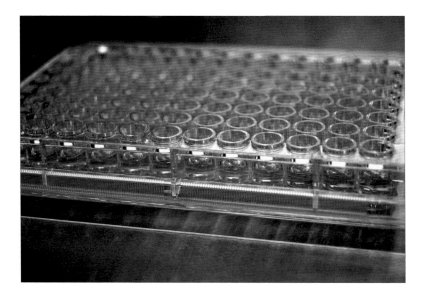

The MIMIC® system, originally developed by VaxDesign, today Sanofi Pasteur VaxDesign Corp., enables researchers to test the immune response to an experimental vaccine or drug developed to fight natural or engineered viruses or other pathogens. Each well in MIMIC's 96-well plastic plate represents a human immune system complete with B cells, T cells, and dendritic cells, the progeny of blood-forming stem cells. The Defense Advanced Research Projects Agency (DARPA) funded the development of the system in the wake of 9/11. *Photo by Todd Lemoine, courtesy of Sanofi Pasteur VaxDesign Corporation.*

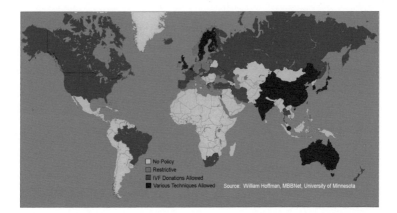

A world map showing which countries have developed national policies supporting human embryonic stem cell research. An earlier version of this map was used during the debate in the U.S. Senate over proposed stem cell research legislation in 2007. Who will lead in the new "Age of Discovery" based on the life sciences? © *Hoffman and Furcht.*

Biopolis One North, Singapore. Biopolis is a multi-billion dollar international research and development center for the biosciences created by the government of Singapore. Biopolis buildings named Nanos, Genome, Helios, Chromos, Proteos, Matrix, Centros, Neuros, and Immunos provide first-rate laboratory facilities for several thosand biomedical scientists. A commentator wrote in 2008 that "the ethos of Biopolis is one of public largesse and liberality rather than the stringent restrictions placed on public funds for embryonic stem cell research in the United States and parts of Western Europe." *Photo credit: Wikimedia Commons.*

The Virgin of the Rocks. Art historians say the painting expresses the kinship Leonardo da Vinci felt with the natural world. For Leonardo, caves represented the mysterious and the unknown. He chose to explore them, first while walking among the hills of his native Tuscany and later in his art, evidence of his relentless curiosity. *Photo credit: Wikimedia Commons.*

class medical research institutions, were among the first states to step up. The second round brought up New York, Maryland, Ohio, Illinois, Oregon, and Washington. But it was California where venture capitalist Robert Swanson and biochemist Herbert Boyer launched the field of biotechnology in 1976 with Genentech, that led the way.

CALIFORNIA

"The embryonic stem cell debate is dead," wrote bioethicists David Magnus and Arthur Caplan in December 2004, a month after the reelection of President George W. Bush. "Long live the stem cell debate." That pronouncement followed in the wake of Proposition 71, California's $3 billion bonding initiative for stem cell research. It's too soon to know whether the results of the referendum mark even the beginning of the end of the debate, but it certainly changed the game.

California was already taking steps to change the game in 2002 when it thumbed its nose at federal restrictions to become the first state to endorse funds for embryonic stem cell research while banning human cloning for reproduction. Then governor Gray Davis said that with its world-class universities, top-flight researchers, and thriving biomedical industry, the Golden State would be a magnet for "the best and the brightest" researchers and halt the migration of scientists to other countries where stem cell research is allowed. "Along with these advances comes responsibilities," added Davis's successor, Arnold Schwarzenegger, in an interview. "There will have to be laws and restrictions created to stop people from misusing the new technology." Schwarzenegger knew that California's economic future rests on smart people and the system of education that supplies them. The state's universities and research institutions claim more Nobel laureates than any state or country in the world. With that as a background, it was not surprising when Schwarzenegger gave the state's stem cell institute a $150 million loan the day after President Bush vetoed a bill to ease federal restrictions on embryonic stem cell research in July 2006.

California's reputation for personal wealth gave hope to advocates of embryonic stem cell research with an eye to moving the domestic research agenda forward with private funds. California is home to more millionaires per capita than any other state; one-fifth of the country's billionaires live in California and represent a combined net worth of more than $100 billion in 2008. One of the first salvos of the creative financing barrage was launched by the University of California at San Francisco, which recruited retired Intel chair and CEO Andy Grove to lead a multimillion-dollar fund-raising campaign for the institution's stem cell program, to which Grove provided a lead gift of $5 million. A coalition in Southern California, organized through the Burnham Institute in La Jolla, put in place an infrastructure of lab networks to share stem cell lines and biomaterials. But it was the ballot drive known as the California Stem Cell Research and Cures Initiative that really got the ball rolling.

In 2004, a coalition of scientists, private financiers, and patient-advocacy groups—particularly wealthy Californians who have children with juvenile diabetes or suffer from it themselves—organized to draft a $3 billion, ten-year general obligation bond issue and put it before voters in the November 2004 referendum. Proposition 71, dubbed a "scientific secession" by *Wall Street Journal* reporters, was voted in after proponents outraised opponents 10–1. Prop 71 appropriates some $300 million per year to support stem cell research. "A key factor: Nearly 85 percent of Californians have a family member or close acquaintance with one of five conditions—Alzheimer's, diabetes, heart disease, Parkinson's or spinal-cord injury—that potentially could be treated with stem cells," observed *The Wall Street Journal*'s Antonio Regaldo and Michael Waldholz. The funds are being distributed by a California Institute for Regenerative Medicine, with oversight by independent citizens' committees drawn from universities and research institutions and headed by Prop 71 campaign leader Robert Klein and, until 2008 when he stepped down, retired scientist and biotech executive Edward Penhoet. The institute models itself after the National Institutes of Health in the way it reviews grant proposals—through a competitive process based on scientific merit.

How did Prop 71 happen? The progress in stem cell research being made in several Asian countries was difficult for California's scientific luminaries to watch from afar. Nobel laureates Paul Berg of Stanford University and David Baltimore, then president of Caltech, plus Stanford's Irving Weissman and former Stanford president Donald Kennedy, then editor in chief of *Science*, weighed in through a commentary in the *San Jose Mercury News* in 2004. "The national debate over human embryonic stem cell research—one that has pitted religious objections against the promise of major scientific and therapeutic advances—has been reawakened by a dramatic advance that could have been made in the United States, but wasn't," they wrote, referring to the report that South Korean scientist Hwang Woo-suk and his team had successfully created an embryonic stem cell line using the SCNT method (which turned out to be faked research). They wished Prop 71 success both as a possible alternative to the federal impasse and as a wake-up call to its consequences in a world where countries like China, Singapore, and South Korea are more than happy to pick up the slack from sagging U.S. science.

The state's academic science stars received firm backing from its globally leading biotechnology industry and city-region economic development agencies as well as disease group advocates. Money poured in from Hollywood and Silicon Valley but also from out of state. Proponents had just a slight edge until Schwarzenegger publicly endorsed Prop 71, putting him at odds with the state's Republican Party. "Californians believe in gravity, not to mention stem cell research. We don't stick science and innovation in the back seat while religion drives the car," wrote *Los Angeles Times* columnist Patti Morrison.

Proposition 71 established a state constitutional right to pursue stem cell research, including therapeutic cloning, but prohibits funding of human reproductive cloning research. In the months following passage of the referendum, as the California Institute for Regenerative Medicine (CIRM) was being set up, Prop 71 organizers were already mounting a campaign to prevent a bill in Congress that would outlaw all forms of cloning. Meanwhile, the location for the institute unleashed a furious round of bids from nearly a dozen cities to host it, a competition reminiscent of that for hosting the Olympic games.

By early 2006, after months of wrangling with legislators and citizens' groups over conflict of interest, intellectual property, confidentiality, and oversight issues, the CIRM found itself in court. It was challenged by taxpayer groups and research opponents representing both ends of the political spectrum. The plaintiffs argued that Prop 71 violated the California constitution by not providing sufficient oversight for the institute. Supporters of the initiative dismissed the challenge as an effort to "sabotage" in court what California voters had overwhelmingly approved at the ballot box. But the tactic was having the desirable effect in the eyes of opponents. The lawsuits prevented the institute from selling any of the $3 billion in bonds because Wall Street investors were afraid to buy them until the legal questions were resolved. Even with the favorable court ruling, officials from the institute said it could take considerable time—many months—for the appeals process to run its course and for bond money to start flowing. One stem cell company, Massachusetts-based Advanced Cell Technology, decided to gamble by opening a ten-thousand-square-foot stem cell lab in Alameda as the CIRM was in court in nearby Hayward.

In May 2007, two and a half years after passage of Prop 71, California's Supreme Court removed the last roadblock to issuing $3 billion in bonds. By then, the CIRM had already approved $160 million in research grants, most for embryonic stem cell research. Governor Schwarzenegger hailed the court's decision, issuing a statement that it "reaffirms voters will to keep California on the forefront of embryonic stem cell research." Even as he spoke, a massive global recession was taking shape that would virtually bankrupt the state, raise questions about the viability of its initiative referendum system, and fuel the political backlash against the massive state investment in stem cell research.

By the end of 2010, seven of twelve CIRM-funded major laboratories and facilities had opened their doors (UC Berkeley, UC Davis, UC Irvine, USC, Stanford, UCLA, and UCSF). The $271 million it provided leveraged more than $800 million in private and institutional funding, according to CIRM. The institute funded an independent economic impact study for the period 2006–2014 that showed CIRM grants totaling $1.1 billion created nearly twenty-five thousand research and construction jobs and generated $200 million in new tax revenue through 2014. CIRM also collaborated with state education officials to develop a model curriculum for the field of regenerative medicine to be offered in public schools following enactment in 2009 of the California Stem Cell and Biotechnology Education and Workforce Development Act.

CIRM funding has been important to progress in embryonic stem cell research given the restrictions in federal funding for the research during the Bush administration and policy uncertainty following court decisions during the Obama administration. Some critics are concerned about heavy CIRM board representation by members of research institutions that stand to benefit from its facilities and research grants. Other critics assert that CIRM must become more industry friendly to ensure successful clinical translation of research advances. From its inception CIRM has seen as its mission to move boldly in translating basic science into therapies. Patient advocacy groups continue to exert steady pressure to ensure that CIRM does not settle into a recumbent world of routine laboratory research without generating therapies that help people suffering from disease and disability. In the end, it is the availability of clinical applications that will be the ultimate proof of whether the unique and monumental CIRM experiment is a success.

NEW JERSEY

California was the first state to approve public funds for embryonic stem cell research. New Jersey was the first state to actually use such funds for stem cell research. The person who made it happen was Wise Young, chair of cell biology and neuroscience and director of the Keck Center for Collaborative Neuroscience at Rutgers University in New Jersey. *Time* named Young "America's Best" in the field of spinal cord injury research in 2002, in part for helping to develop a high-dose steroid treatment as the first effective therapy for spinal cord injuries.

Inspired by a visit to his homeland of Japan, where he saw promising human trials in spinal cord injury research, Young promised actor Christopher Reeve, with whom Young had worked to explore stem cell therapy in spinal cord injury, that he would push for the possibilities of U.S. science despite the challenges often expressed by Reeve himself: Why is there government support for all forms of stem cell research in Britain, Sweden, Belgium, Switzerland, Israel, Singapore, and Japan, but not in the United States, Germany, France, and Italy? Why do allied countries that share many cultural, religious, and ethical standards have opposing and divisive public policy on this research?

Through Young's efforts, and those of New Jersey's disability groups and the pharmaceutical industry (the state is home to Big Pharma companies like Merck, Schering-Plough, Johnson & Johnson, and Bristol Myers Squibb), the legislature passed a bill in 2004 authorizing stem cell research to be done in New Jersey. Governor James McGreevy then proposed that $6.5 million in state funds be used to create a research institute as part of a five-year, $50 million public-private initiative. McGreevy signed legislation in May 2004 establishing the nation's first state-supported stem cell research facility. In late 2005, New Jersey became the first state to use public money to fund human adult and embryonic stem cell research, announcing $5 million in grants to be split among seventeen research projects. A year later, the state legislature approved spending $270 million on stem cell research facilities in Camden, Newark, and New Brunswick. But the state's grand plans to compete with California suffered a setback in 2007 when New Jersey voters rejected Governor Jon Corzine's bid to borrow $450 million through a bond measure to fund stem cell research over ten years. Christopher Scott, director of Stanford University's Program on Stem Cells and Society, observed that the proresearch community in New Jersey was not as diverse or as well organized as in California, which energized Hollywood celebrities and biotechnology executives as well as scientists and patient advocacy organizations.

The Stem Cell Institute of New Jersey, a joint venture between Rutgers and the University of Medicine and Dentistry of New Jersey-Robert Wood Johnson Medical School, broke ground on an eighteen-story facility in 2007. Despite a number of slowdowns and setbacks, the facility, with its Christopher Reeve Pavilion, is scheduled to be completed in 2012. Meanwhile, Wise Young is pushing hard for human clinical trials to test spinal cord injury therapies in the United States. They will build on trials that Young established in Hong Kong and China through the China Spinal Cord Injury Network (ChinaSCINet) he formed in 2008. China is estimated to have some 40 percent of the world's spinal cord injury patients with approximately sixty thousand new cases each year, mainly from automobile, construction, and mining accidents. The Chinese trials use umbilical cord blood cells and the chemical lithium that stimulates neurons to grow, both in combination with rigorous physical therapy. American patients have asked him how they could go to China to join the trials. "How far have we declined in this country that we have to send people to China to participate in clinical trials of therapies developed in the U.S.?" Young asked in an interview with *NewJerseyNewsroom.com*. "It isn't that umbilical cord blood cells and lithium are at all controversial. The only obstacle is money." If the ChinaSCINet trials demonstrate that umbilical cord blood stem cell transplants help patients with spinal-cord injury, Young will ask the U.S. Food and Drug Administration (FDA) for approval to undertake a phase 3 trial within the North

American Spinal Cord Injury Network, which Young established in 2009. "Regardless of the outcome of the clinical trials, it will be a significant achievement to demonstrate that this can be done in China," stem cell researcher Stephen Minger told *Lancet Neurology*. "To have a stem cell trial approved by the FDA based on studies in China would be rather extraordinary."

MASSACHUSETTS

During the first decade of the 2000s, Massachusetts sought to cement its reputation as a magnet for biomedical research and entrepreneurship. By early 2004, the state had started to make its own ripples in what was described as a "tidal wave" of state initiatives supporting stem cell research. Massachusetts legislators resurrected a bill calling for the creation of a trust fund that would provide not only tax incentive credits to entrepreneurial firms in the field, but also grants to nonprofit and for-profit research organizations doing embryonic stem cell research.

Originally part of an economic stimulus bill approved by the state senate, the stem cell proposal had been removed during the previous session due to effective opposition by state Senate Speaker Thomas Finneran and Governor Mitt Romney. By 2005, however, the mood in Massachusetts had changed. The future of the biotech mecca was perceived to be at risk in light of California's passage of Prop 71 and the anticipated "stem cell gold rush" out West. Finneran found himself head of the Massachusetts Biotechnology Association as if experiencing a conversion on the road to Damascus. Governor and future Republican presidential candidate Romney, on the other hand, stated that he would support criminal penalties for researchers who create human embryos through therapeutic cloning. A couple of months later, Harvard provost Steven Hyman announced that the university had approved the cloning of human cells to make embryonic stem cells, putting the institution directly in Romney's crosshairs. The then Harvard President Lawrence Summers had been positioning the university to move boldly into the era of the life sciences. In his inaugural address in 2001, Summers noted that "we live in a society, and dare I say a university, where few would admit—and none would admit proudly—to not having read any plays by Shakespeare or to not knowing the meaning of the categorical imperative, but where it is all too common and all too acceptable not to know a gene from a chromosome or the meaning of exponential growth. . . . Part of our task will be to assure that all who graduate from this place are equipped to comprehend, to master, to work with, the scientific developments that are transforming the world in which we will all work and live."

At the forefront of those developments is the Harvard Stem Cell Institute (HSCI). Constituting its research focus are diabetes, heart disease, kidney disease, neurodegenerative diseases, blood and immune diseases, and cancer. Stem cell scientist George Daley told reporters in 2004, the year the center opened, that "Harvard has the resources, Harvard has the breadth, and frankly, Harvard has the responsibility to take up the slack that the government is leaving." HSCI created an iPS Core Facility that serves as a repository for reprogrammed adult cells produced by HSCI scientists and "functions as a laboratory to produce disease-specific lines for sharing with the HSCI and broader research community." But its key goal is to bring together the growing number of researchers from the university and its powerful constellation of affiliated hospitals.

One of those researchers is Douglas Melton, a Harvard biologist who codirects the institute. Melton has more than a professional interest in making embryonic stem cells work for medicine. He has two children with type 1 diabetes. In his mind, stem cell research is the only way to meet the goal of finding a cure for diabetes. With several hundred leftover embryos obtained with donor consent from Boston IVF, one of the nation's leading infertility clinics, Melton created seventeen new lines of embryonic stem cells. He made the lines available in 2004 to any scientist who wants them free of charge for noncommercial use.

In 2007, Melton was named one of *Time* magazine's "Time 100," a list of "100 men and women whose power, talent or moral example is transforming the world." Melton compares the power of stem cells in medicine to the power of the transistor in communications. "I believe the twenty-first century will be the century of cells and of stem cells," he told the Harvard University *Gazette*. As if taking his cue from Melton, Massachusetts Governor Deval Patrick told attendees of the BIO 2007 meeting in Boston that he would push for a ten-year $1 billion investment in biotechnology: $500 million in state bonds for capital investment and $500 million in research grants and tax credits. The centerpiece of his proposal is the creation of a Massachusetts Stem Cell Bank at the University of Massachusetts. "You cannot be in the company of someone you love, powerless to help them, without appreciating the vital importance of stem cell research and other biomedical breakthroughs," Patrick and state senate president Therese Murray wrote in the *Boston Globe*.

The University of Massachusetts Human Stem Cell Bank and Registry opened in early 2011 with seven stem cell lines made available for worldwide distribution to researchers working on discovering new therapeutic treatments for diseases such as cancer, juvenile diabetes, Alzheimer's, and Parkinson's. The initial lines made available—five human embryonic stem cells lines and two induced pluripotent stem cell lines—were developed in the laboratory of George Daley at Children's Hospital in Boston. The facility plans to bank many additional stem cell lines including lines from the Harvard Stem Cell Institute, which is creating disease-specific induced pluripotent stem cells lines, or "diseases in a dish," at a healthy clip. Such lines will be invaluable to researchers tracking how disease originates and how it can be treated and prevented.

WISCONSIN

The "war between the states," in contemporary parlance, normally refers to state economic development entities trying to lure businesses to their state in the interest of growing jobs. Science is not immune from such tugging and pulling. When he announced the availability of the human embryonic stem cell lines he developed, Harvard's Douglas Melton fired a volley westward over the Great Lakes to WiCell, the U.S. fountainhead of embryonic stem cells. WiCell is an affiliate of WARF, the Wisconsin Alumni Research Foundation, an arm of the University of Wisconsin and the largest holder of federally approved and available stem cell lines in the country. Melton was unhappy with embryonic stem cell lines he received from WiCell, saying they were unreliable and pricey at $5,000. His criticism prompted a swift rejoinder from WiCell, complete with follow-up polling data of WiCell customers.

Whatever the merit of Melton's charges, WiCell was firmly center stage in the stem cell research world and had been since August 2001, when its scientific director, James Thomson, appeared on the cover of *Time* as "the man who brought you stem cells." Thomson, a developmental biologist at the University of Wisconsin's primate center, pioneered the creation of human embryonic stem cells in 1998. WARF owns a U.S. patent that covers both the method of isolating embryonic stem cells and the cells themselves, a patent the U.S. Patent and Trademark Office decided to review in 2006 after complaints that it was overly broad. Then in May 2007, the U.S. Supreme Court issued a patent decision that strengthened challenges to the WARF stem cell patents, at least in the view of the two California nonprofits fighting to loosen WARF's hold on the cells and methods for isolating them.

Charges that WARF's patents were holding up research progress had an effect. When Thomson's research team reported that it had successfully reprogrammed skin cells and from them created pluripotent stem cells, WARF managing director Carl Gulbrandsen told the press that his office would not require researchers to obtain a license this time to use the technology once the patent application for the reprogramming method was approved. "They can do it in their own lab," he told *The New York Times.* "They don't have to tell me about it, and I don't really have to know." Of course, that depends on the fate of Wisconsin's patent application. How will intellectual property credit for the reprogramming methodology be divided up between the Wisconsin and Japanese claims? In the view of bioethicist Arthur Caplan, "A huge and probably nasty battle is likely to quickly erupt over priority and patenting in this area. Given the potential payoffs involved, the fight will likely keep lots of lawyers busy for years to come."

Wisconsin wants to be a winner in the stem cell economic development sweepstakes. During his term as governor, Tommy Thompson was a big supporter of stem cell research in the state. When Thompson became Secretary of Health and Human Services in George W. Bush's first administration, he spoke several times with James Thomson while advising the president on policy. Whether the stem cell policy promulgated from Crawford, Texas, that day in August 2001 actually squared with Tommy Thompson's libertarian instincts, he didn't say. Following his departure from the Bush administration in 2005, however, Tommy Thompson made it clear that he is a strong advocate of stem cell research.

It is no little irony that both the secretary and the star figured in a federal lawsuit filed in May 2001. In *Thomson v. Thompson,* the plaintiffs claimed that the Bush administration illegally withheld federal funding for stem cell research, funding that could save lives. Christopher Reeve and seven scientists filed the suit. It was withdrawn one week after Bush spoke on national television outlining his administration's policy. Back home in Wisconsin, the governor turned secretary was credited with putting the state and its university on the map by persuading his boss to permit federal funding of research on embryonic stem cells already in existence, of which at the time Wisconsin had more than anybody else.

In 2004 the University of Wisconsin instituted a stem cell program that coordinates embryonic and adult stem cell research and education activities across the Madison campus. It also launched a "cluster hiring initiative" to attract top talent in stem cell biology, tissue engineering, and regenerative medicine. The Wisconsin State Assembly, meanwhile, was divided on the subject. In August 2001, the legislature passed a joint resolution praising Thomson's pioneering research, stating that it had "significant potential" to improve public health and economic development in Wisconsin and calling for the National Institutes of Health to designate the University of Wisconsin as a National Center of Excellence in Embryonic Stem Cell Research. At about the same time, the state legislature introduced the Human Embryo Protection Act which, had it passed, would have outlawed embryonic stem cell research in the state and also stem cell distribution by WiCell. Thomson warned legislators that if Wisconsin is perceived as antiscience, it could forget about becoming a biotechnology center.

Indeed, the highly visible moves by California, New Jersey, and Harvard University caught the notice of Wisconsin's academic and political leaders. Two weeks after passage of Proposition 71, Wisconsin governor Jim Doyle announced plans to invest nearly $750 million to support human embryonic stem cell research and perhaps to provide political cover for other leading biomedical research fields. He said the proposal would build on the nearly $1 billion the state has spent on high-tech facilities during the previous fifteen years. Wisconsin's status as the birthplace of human embryonic stem cell lines was a major issue in the state's 2006 gubernatorial race and helped Doyle get reelected over his Republican opponent Mark Green. Green lost despite the support of the former four-term governor and Health and Human Services Secretary Tommy Thompson.

A group of Wisconsin business leaders, most of them Republicans, took a page from California and New Jersey and set up a state foundation and fundraising program. One of them put the matter plainly, as it has been put plainly in other states where restrictive laws are being challenged: With Wisconsin and other states hemorrhaging jobs in the traditional industries of farming and manufacturing, "it can't afford to pass on the employment opportunities offered by the new technologies in both the research and health care sectors."

Indeed, with manufacturing moving to China, information services to India, and agriculture to Brazil and Argentina, lawmakers in upper Midwest states who turn their backs on research and innovation in the emerging biosciences will be called on the carpet when the standard of living of their constituents begins to slip, as in some cases it has already. Their job is made easier with scientists such as James Thomson on board. Besides providing innovative ideas and intellectual property, superstars like Thomson invariably attract a large supporting cast of scientific and entrepreneurial talent.

"The political battle in Wisconsin over stem cell research ended with the election of a governor who supported it," Governor Doyle told an audience of more than a thousand people gathered at the World Stem Cell Summit in Madison in September 2008. "Nationally, the same thing can happen. I want to make sure that we have a president who rescinds the restriction on NIH funding as one of his first acts." Doyle was granted his wish, but less than two years later, with the election of Republican Scott Walker to succeed him, stem cell politics returned to the Badger State. During his campaign, Walker said scientists "have shown us (that) the greater possibilities, the real science movement, has been with adult stem cell research. It has not been with embryonic," adding "that's not a political statement; that's a statement of scientific fact out there." The Wisconsin edition of the respected PolitiFact fact-checking service rated Walker's claim as "false" based on information from the International Society for Stem Cell Research, the National Academies of Science, and the Medical College of Wisconsin. But in the political arena, misinformation has its purpose. As the new governor sought to restructure the relationship of public employee unions with the state of Wisconsin, stem cell researchers on the Madison campus waited to see whether he would seek to alter the state's relationship with an enterprise that had put Wisconsin on the scientific map.

COMPETITORS ABROAD

The economic competitors of the United States are no longer operating just from across the river or across the country but from across oceans and continents. Facilitating the flow of capital and growth of emerging markets is information technology, "a technological revolution of momentous and uncertain consequences," wrote Daniel Yergin and Joseph Stanislaw in *The Commanding Heights*. "Information technology—through computers—is creating a 'woven world' by promoting communication, coordination, integration, and contact at a pace and scale of change that far outrun the ability of any government to manage." Some competitors in the biosciences are taking advantage of the Internet to foster academic-industry linkage despite their smaller economies and bioscience infrastructure relative to the United States. Less advanced nations may target their information technology resources to gain a competitive advantage in the biosciences and other fields. They see electronic images of what wealth enables. They may be willing to make personal sacrifices and dodge moral deadlocks as they learn how to play the game. It is a game they intend to win. "What is emerging is a global science system in which the U.S. will be one player among many," said Caroline Wagner of Penn State University at the 2011 annual meeting of the American Association for the Advancement of Science in Washington, D.C.

As we will see, fast-growing Asian countries are on the move in the sciences including regenerative medicine. They are building up their educational and research infrastructure with massive investments. But in the short term it is with large and progressive clusters of innovation in countries like the United Kingdom, where great minds from Sir Isaac Newton to Charles Darwin to Francis Crick to Stephen Hawking have ensured the country's sterling academic reputation, that will be most competitive to the United States as the stem cell research and therapy fields evolve.

ENGLAND

In the West, the United Kingdom has had more than its share of biomedical research breakthroughs. In the last six decades alone, its scientists have discovered the structure of DNA; isolated the mouse embryonic stem cell, the cornerstone of mammalian stem cell biology; and cloned Dolly the sheep. Thirty years ago, the United Kingdom also set the stage for the emergence of human stem cell research by pioneering in vitro fertilization. That feat led not only to the world's first test tube baby and more than four million like her but also to hundreds of thousands of leftover embryos—embryos such as those that Wisconsin's James Thomson used to find, extract, and cultivate human embryonic stem cells.

The United Kingdom is using its history of stem cell research to position itself as the research hub of the future. Addressing the Royal Society in early 2005, then Foreign Secretary Jack Straw vowed that Britain would not be deterred from pursing stem cell research after debating questions posed by embryo research for two decades and being one of the first countries to ban human reproductive cloning, which it did in 2001. "But we were equally clear then—and remain so now—that therapeutic cloning and stem cell research, in a properly-regulated environment, holds enormous promise of new treatments for diseases which kill many millions of people every year," he said, adding that, on the question of therapeutic cloning, "our specialist science officers and our diplomats around the world have helped to shift international opinion and get the idea of a UN vote on such a ban taken off the agenda."

It is no accident that the U.K. Trade and Investment office, which for years has been preoccupied with improving the country's international competitive stature, is working closely with public and private enterprises to promote stem cell research. During his long tenure as minister for science and innovation, Lord David Sainsbury of Turville made a commitment "to help create a climate in which stem cell research can flourish." Perhaps no public official trumpeted stem cells like Sainsbury, a past chair of Sainsbury PLC, one of the world's largest food retailers.

Backing the overall effort is a weighty lineup of the British scientific and health care establishments. In 2002, for instance, the U.K. Stem Cell Bank was set up under the National Institute for Biological Standards and Control, with about $5 million in funding from the Medical Research Council and the Biotechnology and Biological Sciences Research Council. The bank is a major repository for stem cells derived from adult, fetal, and embryonic tissues that is open to academics and industrialists everywhere. In the words of the moral philosopher Baroness Onora O'Neil of Bengarve, the bank was formed to make the United Kingdom the last choice rather that the first for someone wanting "to carry out irresponsible research on embryonic stem cells."

In December 2005, British Chancellor Gordon Brown announced that $190 million would be made available for U.K. stem cell research over two years to help develop treatments for incurable illnesses and medical conditions. Some months earlier, a private group that includes Virgin Airlines founder and billionaire Sir Richard Branson, the president of the Royal Society, and a former U.K. chief scientist proposed establishing a $190 million U.K. Stem Cell Foundation. In his final speech to the Labour Party in late 2006, Prime Minister Tony Blair observed that "if America does not want stem cell research—we do." Just weeks earlier, he had visited Silicon Valley to build ties with California's stem cell researchers and biotech businesses.

Yet there's a critical subtext of the stem cell story, something apart from the hope for better treatments and clinical outcomes.

That subtext is really about the connection between science, economics, self-image, and power. It is about the race for international leadership. "Biotechnology is the next wave of the knowledge economy and I want Britain to become its European hub," Blair told the Royal Society, Britain's national academy of science, in 2000. He backed up his words with money, funneling hundreds of millions of British pounds into government-funded R & D to go with a host of financial and tax incentives and an aggressive international partnership and marketing program.

Competitive pressure and election-year politics in the United States eventually compelled the Bush administration to respond. In July 2004, as George W. Bush was gearing up his reelection campaign, Health and Human Services Secretary Tommy Thompson announced two new U.S. initiatives related to human embryonic stem cell research. He said the NIH would create a National Embryonic Stem Cell Bank that "will consolidate many of the cell lines eligible for funding in one location, reduce the costs researchers pay for the cells, ensure uniform quality control, and further our knowledge about the cells themselves." The NIH also would spend $18 million to create at least three new Centers of Excellence for Translational Stem Cell Research, "with the goal of exploiting new discoveries in basic embryonic and stem cell biology." Thompson's announcement "will not stop the exodus of stem cell research from the United States to Europe, particularly Great Britain," *The Sacramento Bee* editorialized.

But British governmental enthusiasm for supporting stem cell research that characterized most of the 2000s changed by the decade's end. The global economic downturn that began with the financial crisis in 2008 helped to bring a conservative government to Britain after nearly two decades of domination by the Labour Party. Prime Minister David Cameron instituted a series of budget austerity measures soon after his election in 2010. Sir Richard Sykes, the chairman of the U.K. Stem Cell Foundation, warned that Britain's leadership position was at stake. "The U.K. is without doubt a leader in stem cell research," he told *The Telegraph*. "We need to translate and commercialise stem cell therapies, or scientists will move away."

Although governmental support for stem cell research reaches universities and research centers located throughout the country, southeast England is key to its global standing. The region that includes and surrounds Cambridge University is Britain's leading claim to being a global competitor in biotechnology, genomics, and stem cell research. Its prestigious Sanger Institute, funded by the Wellcome Trust, was a major partner in the Human Genome Project. The noted scientist Austin Smith, codiscoverer of the stem cell gene *nanog* named for the Celtic "Land of Forever Young," was appointed director of Cambridge University's Wellcome Trust Centre for Stem Cell Research in 2006. The stem cell company ReNeuron, located in Surrey near London, launched a two-year human clinical trial for treatment of stroke with its genetically engineered neural stem cell therapy in 2010. Treatment for stroke coupled with lost productivity costs Britain more than $8 billion annually.

To build a biosciences hub, it takes more than just brains and talent and a favorable regulatory regime. The biotechnology push in the United Kingdom is being held back, as it is everywhere else, by an investor mind-set that is reluctant to pull the trigger for research that does not have near-term commercial potential. To see where the stem cell–based industries of the future will take root, follow the money. The trail is leading to Asia where educated and energetic youth, growing scientific sophistication, government support, and private investment are combining in a perfect storm of competition to make Asia, over time, a global contender in the emerging biosciences.

SINGAPORE

No Asian country has adopted Western technology and stem cell competitiveness as energetically as the thriving city-state of Singapore, a financial and microelectronics manufacturing center and one of the juggernauts of oriental capitalism. Singapore packs more than five million people—an ethnic and religious mix of Chinese, Malays, Indians, Buddhists, Muslims, Christians, and Hindus—into an area of just 250 square miles. Several million square feet of that space is home to Biopolis, the government's ambitious plan to make Singapore a global hub for biomedical research and development, with stem cell research at its vital core. With its generous research grants, tax breaks, and numerous other governmental incentives worth many millions, Biopolis is the central hub. A $3.5 billion investment, Biopolis is located close to Singapore National University and Singapore National Hospital, where Sri Lankan native Ariff Bongso, an IVF specialist trained in Canada, is the local stem cell star. Phase 2 of Biopolis, which opened in 2006, consists of two city blocks named Neuros and Immunos where neuroscience and immunology reign.

Singapore has always relied on the ingenuity and brainpower of its people even if its government is the controlling kind. Recognizing that success in biotechnology, with its potentially big payoff, is a long-term proposition compared to Internet start-ups, Singapore committed nearly $1 billion in 2001 to a five-year plan designed to build the biosciences in the city-state. It funneled more biology into school curricula, endowed more scientific scholarships, and funded more entrepreneurial competitions in the field. Philip

Yeo, Singapore's top science and technology economic planner at the time, began aggressively recruiting life and biomedical scientists to complement its abundance of engineers. Pharmaceutical and biotechnology firms looking to set up operations or expand existing operations globally and in Asia are being given the tax-incentives red carpet treatment. Singapore's biomedical manufacturing output more than tripled from about $5 billion in 2000 to more than $16.2 billion in 2009. The biomedical sector now represents 10 percent of the country's manufacturing output, up from 4 percent in 2000.

In 2003, Biopolis hosted more than five hundred physicians and scientists in an event billed as "the world's biggest stem cell conference." The conference, which coincided with the grand opening of the $500 million research park, was organized not by Singapore's health ministry but by A*Star, the Agency for Science, Technology, and Research, chaired by economic development minister Yeo. Sir George Radda of Oxford, who chaired the conference, had just stepped down as chief executive of the UK's Medical Research Council, where he led an initiative for international cooperation with the endpoint being a full-fledged Human Genome Project for stem cells. To guide its scientific development, A*Star enlisted British Nobel laureate Sydney Brenner, who works at the Salk Institute in La Jolla, California. Biopolis can provide first-rate laboratory facilities to some two thousand scientists. Its openness to stem cell research is a calculated strategy to lure frustrated scientists in the United States and Europe to the intellectually boundary-free city-state.

The new Genomics Institute of Singapore is another piece of the strategy to bring Western scientists to the Pacific Rim. "Open-source" initiatives and collaborations, such as the Asia-Pacific International Molecular Biology Network, may do for Asia what the

European Organization for Nuclear Research and the European Molecular Biology Organization have done for Europe: stimulate regional cooperation. They could help Singapore shed its image as a facsimile of a Big Brother regime and therefore lure Westerners to deliver its biofuture.

But if Singapore were going to compete with the United States and Europe, it knew it needed more than a marketing expansion strategy. It needed the knowledge of American academic medical institutions. With investment from Singapore's biomedical fund, the National University of Singapore formed a partnership in 2005 with Duke University School of Medicine to establish the Duke-NUS Graduate Medical School of Singapore. Duke-NUS launched strategic research programs in cancer and stem cell biology, neurobehavioral disorders, infectious diseases, cardiovascular and metabolic disorders, and health services research. In 2010, Duke-NUS researchers reported that they had identified a novel feedback mechanism that controls the delicate balance of brain stem cells and normally prevents them from becoming diseased. It was another sign that the brainpower responsible for Singapore's rapid economic rise was entering a new phase.

SOUTH KOREA

Korea is called the "Land of the Morning Calm." But the Republic of South Korea was anything but calm during the last two decades of the twentieth century. Thanks to huge investments in science and math, its economy climbed to fifteenth in world rankings (nominal gross domestic product, International Monetary Fund, 2009). It became, as it is today, the country with the greatest penetration of broadband technology.

Such was the competitive soil that yielded the one-time Korean cloning superstar Hwang Woo-suk. As the lead scientist in a landmark study published in February 2004, Hwang claimed to have made the first stem cell line from a cloned human blastocyst, the earliest stage of embryonic development. In a subsequent study published in 2005, Hwang made an even bolder claim: he and his group reported that they had established a series of stem cell lines from cloned blastocysts created by using DNA from persons with various diseases. Hwang became a national hero, the recipient of huge government resources and awards and the object of massive media attention, including being listed by *Time* among the "100 people who shape our world."

By late fall 2005, Hwang's triumph had turned to tragedy, as much of the data behind his assertions revealed multiple layers of deceit. Egg donors who had been previously identified as volunteers were in fact employees subordinate to Hwang's authority. When Hwang's academic institution, Seoul National University, investigated the claims of the work after the ethical breaches first came to light, an investigatory panel found that the so-called clones were not clones at all and were in fact genetically different. What in retrospect might be seen as a rush to judgment by *Science,* the journal that published the paper, definitive proof was never requested in advance of publication. In addition, the patient-tailored stem cells should have been genetically identical to the cells of the patients themselves, which they were not.

That Korea entered the global competition stage with Hwang's fabrication of much or all of his human cloning data was a failure for all of Korea. The country's political and business leadership had pushed hard for science and math education throughout its primary and secondary school systems during the 1980s and 1990s, and the investment propelled South Korea into a leadership role in the Asian economy. The rapid ascendance of Korean business, electronics firms like Samsung and others that grew rapidly at the expense of Sony and other Japanese competitors, is evidence of aggressive but ethical business practices and great visionary leadership.

In 2006, the South Korean government launched Bio-Vision 2016, a multibillion dollar infrastructure plan to establish a biotechnology-based economy in a decade. Stem cell research is a key component of the plan. Yet the South Korean government put stem cell research using human eggs on hold for three years following the Hwang affair, until 2009 when it resumed, subject to strict oversight. The same year the government tripled its funding for embryonic stem cell research to $32 million a year through 2014 and announced plans to build a national stem cell bank to provide stem cell lines for use in research and therapy. In the private sector, a number of small companies focusing on cell replacement therapies using cord blood stem cells in which South Korea has significant research strength have proceeded to clinical trials following regulatory approval.

Korean stem cell scientists collaborate widely with their colleagues in the West. A research team funded by the Korean Ministry of Health and Welfare joined with American and Canadian teams to develop a novel technique for reprogramming cells that could be used for regenerating tissue in heart disease and stroke. As they reported in 2011, by adding a single gene the teams instructed neural progenitor cells to self-renew in a laboratory dish. Then they introduced these cells into a rodent stroke model where the cells started differentiating and improved brain function in the animals. Seoul's successful hosting of the Stem Cells Asia 2010 international conference was another step in collaboration and recovery from the Hwang affair. The commemorative postage stamp the government issued in Hwang's honor in 2005 showing a man rising from his wheelchair then running, jumping, and hugging his wife—all made possible by stem cell research— proved to be a huge embarrassment.

CHINA

China, during the Ming Dynasty, was the world technological leader when the Renaissance was just getting under way in Europe. China had gunpowder, paper, printing, cast iron, the compass, and sea power embodied in its "treasure fleets" under the command of Zheng He, whose voyages of discovery to Arabia and Africa made China the global superpower of its day just two decades before Leonardo da Vinci's birth in 1452. Then, writes scientist Jared Diamond in his *Guns, Germs, and Steel*, China experienced a power shift at home and retreated into a cocoon for centuries, as if experiencing its own Dark Ages.

Modern China seems determined to recapture past glory. It has bold aspirations in global capitalism, even if it has relatively fewer scientific facilities and resources than its competitors. The biggest asset that China has is its brainpower, both present and future, for the biotechnology battles that lie ahead. In 2003, *Nature* highlighted Chinese laboratory breakthroughs and government support for regenerative medicine. By the end of the decade, the National Science Foundation's *Science and Engineering Indicators* showed China making large gains in internal research and development investments, as well as in investment by multinational corporations. In 2004, the American medical products company Johnson & Johnson, for example, became a full partner in the $300 million firm Xi'an-Janssen Pharmaceutical, the largest pharmaceutical joint venture in China. China is also making large gains in the number of Chinese nationals earning science and engineering doctorates in the United States.

No country sends more of its students to America for higher education than China. Even after visa restrictions put into place following the terrorist attacks of 9/11, Chinese students continue to seek training at MIT, Stanford, and other top U.S. universities. The great majority of them remain in the United States, but that is likely to change. U.S.-trained Chinese students remain closely tied to their native land, and more of them are returning than in the past, lured by the promise of jobs and government funding. When the "sea turtles," as they are called in China, do return to their homeland, they bring with them knowledge of modern science, biology, agriculture, and engineering that is given free rein and support in Chinese research facilities and corporations.

One who came home is Robert Chunhua Zhao, executive director of the National Center for Stem Cell Research in Beijing. He worked with adult stem cell and bone marrow stem cell research scientist Catherine Verfaillie at the University of Minnesota before returning to China in 1999. His research in the United States was on bone marrow stem cells that can be coaxed into making specific blood cells that can then be transplanted into persons with leukemia after their cancer cells have been destroyed. Zhao told *Nature* that China has the clinical know-how to succeed in the stem cell arena. But a country with a nominal per capita income of some $4,000 cannot hope to extend the benefits of stem cell therapies to most of its citizens, at least not today. Most likely it's the economic payout, power, and influence that come with scientific prestige that China's leaders are seeking in the near term, not just improvements in health care for all of its people.

That China wants to be a player in the most audacious exercise ever of biological self-analysis is what is telling. China was the only developing country to participate in the Human Genome Project, making a significant contribution to the project in decoding the genomes of the silkworm and chicken. True, China's aspirations in biomedicine are dwarfed by a vastly superior biomedical and genomics infrastructure in the West, at least for the time being. With stem cells, however, it may be a different story.

China has established cell banks for blood-forming stem cells and umbilical cord stem cells in Beijing, Shanghai, Sichuan, Guangdong, and Zhejiang. In addition, China's more than fifty ethnic groups and homogeneous populations make it prime real estate for tracking disease genes. "The large population also makes it easier to arrange clinical trials," a survey noted. And in the age of genomics, "targeted clinical trials based on the genomics characteristics of the population will play a key role in health biotechnology."

China's successes in producing transgenic rabbits, goats, and cattle (those that contain genes of another species) and in cloning goats, cattle, and rats give it a competitive foundation for becoming a leader in embryo technology. Its willingness to pursue human-embryo technology gives it an edge as well. "China has probably the most liberal environment for embryo research in the world," wrote Xiangzhong "Jerry" Yang in *Nature*. Yang, who received a PhD from Cornell University in animal science, became an American "celebrity scientist" through a series of cattle cloning successes, as well as a "second generation cloning" effort—the cloning of a clone—that created a healthy black bull. In early 2005, he announced that his Center for Regenerative Biology at the University of Connecticut, in collaboration with the Institute of Zoology of the Chinese Academy of Sciences, was first to create embryonic stem cells from cloned cattle embryos.

Within a couple decades China is expected to become the largest economy in the world. A country of 1.3 billion people with a good educational system and hardworking people, coupled with targeted investments and permissive government policies in biotechnology, is rapidly becoming a global force to be reckoned with. In 2007, China spent $102 billion on research and development compared to $369 billion by the United States, $148 billion by Japan, and $72 billion by Germany. Most of the research in China goes into applied technologies. A headline "U.S. still leads in R&D research, but China closing gap fast" captured the essence of a report published in *R&D Magazine* in 2006. China, with more than a decade of open-market reforms under its belt, is beginning to target its resources, which include a $183 billion trade surplus with the rest of the world as of 2010. Regenerative medicine is one of the technologies in the vicinity of the bull's-eye. It is part of China's ambitious fifteen-year Science and Technology Plan.

There is no substitute for conducting on-the-ground surveys and face-to-face interviews to find out what the state of regenerative medicine actually is in any given country. That is exactly what University of Toronto researchers did in China. They identified and consulted the main research institutions, universities, hospitals, firms, funding organization, regulatory organizations, and policymakers active in this field as well as supply institutions such as cord blood bank, infertility clinic, and animal model facilities. In their report, "Cultivating regenerative medicine in China," published in 2010, they concluded that China "has been able to catapult itself into the forefront of regenerative medicine" thanks to "strategic government funding, permissive regulations, a focus on application and a highly educated workforce built by attracting scientific diaspora trained at the foremost stem cell research centers in the world."

Regenerative medicine research in China is subject of "national pride," wrote the Toronto research team. The Chinese feel that their discoveries "can help move global research toward clinical solutions." Because China is focused on developing treatments and less on basic research in the field, it needs to hone to international regulatory standards for clinical testing of stem cell therapies or risk losing international confidence in its amazing progress in the field. By doing so, China could have the impact on the field the Middle Kingdom's stem cell research community envisions just over the horizon.

PATIENT ADVOCACY GOES GLOBAL

Emerging to have their say in our stem cell future are powerful disease-group lobbies and advocacy organizations, breaking out from within their national boundaries into the global science and technology arena. They use the power of empathy to raise their own money and to leverage the capital of others on a scale that can serve as something of a counterweight to restrictive governmental policies in biomedical science.

Consider the Singapore meeting in October 2003, sponsored by the city-state's Agency for Science, Technology, and Research (A*Star). The disease-group life force at that convention was the New York–based Juvenile Diabetes Research Foundation International. JDRF and Singapore's Biomedical Research Council, an arm of A*Star, signed an agreement to fund stem cell research in Singapore, with each contributing half of the $3 million project. The World Health Organization estimated that diabetes afflicted 171 million people worldwide in 2004 and is projected to reach 366 million by 2030. Approximately three million Americans have type 1 diabetes, the form of the disease that affects children and young adults primarily. Both the United States and Singapore have rapidly growing patient populations, and research advocates in both countries see stem cells as the most promising avenue for eventual treatment success.

Founded in 1970 by parents of children with diabetes, the Juvenile Diabetes Research Foundation is the world's leading charitable funding source of diabetes research. The United States is as generous as it is competitive, and as a result, big money has flowed into JDRF. It has funded more than $1.5 billion throughout its history. That is one of the keys to its success. Its bold mission of being "a diabetes cure enterprise" has, in turn, prompted the organization to move into the strategic investments and international research partnership arenas. JDRF, while based in the United States, is "building a global network of scientists who share our commitment to a cure." In the 1990s, it funded the team that achieved what has been termed the Edmonton Protocol, considered the first reproducible proof-of-concept in using pancreatic islet cell transplants to restore normal blood sugar in people with type 1 diabetes—the road map for a cure for diabetes.

But the protocol can't help most persons with diabetes because of the vast numbers of people with the disease, the short supply of pancreases and their islet cells available for transplant and, in some cultures, the objection to organ donation and transplantation. To deal in any meaningful way with the overwhelming number of diabetic patients, a renewable source of insulin-producing cells is needed. The potential of embryonic stem cells as a potential in diabetes treatment showed up quickly on JDRF's radar screen. By the time President George W. Bush announced his administration's policy on stem cell research, JDRF money was already flowing into James Thomson's stem cell research laboratory in Madison. "We're not waiting," Thomson told the *Milwaukee Journal Sentinel* a month before the announcement. That was the essence of the JDRF message when it launched a $20 million fund-raising effort to support stem cell research in 2002, including research on cells not eligible for public NIH money because they came into existence after August 9, 2001.

JDRF changed its strategy when its scientific review committee found that some of the best stem cell and diabetes research was being done outside U.S. borders. In fiscal year 2002–2003, it spent $3 million on human embryonic stem cell research, most of it going to researchers in Australia, Belgium, Canada, Denmark, Finland, France, Germany, Israel, Italy, Japan, Korea, Norway, Spain, Sweden, Switzerland, Ukraine, and the United Kingdom. In the U.K., it teamed up with the Wellcome Trust, the world's largest medical research charity, to support stem cell research over five years. But it was Asia that caught JDRF's eye. The progress the Asians have made is "astonishing," JDRF's chief scientific officer Robert A. Goldstein told *BusinessWeek* in 2005. As he saw it Asian governments were asking themselves: "Since the U.S. doesn't seem to be taking a lead role, why don't we?"

The trend was clear: America might have the research talent and infrastructure at home, but more and more of the action, including human clinical trials, was moving offshore. The trend is not necessarily caused by but was certainly related to more restrictive government policies in the United States during the Bush administration. The National Institutes of Health spent only about $40 million annually on human embryonic stem cell research until President Barack Obama eased the restrictions in March 2009, a move that resulted in NIH funding for such research growing to $123 million by 2010. As we saw in the previous chapter, that year, a court challenge based on the Dickey-Wicker Amendment outlawing research on human embryos put all federal funding for human embryonic stem cell research at risk. The policy tug-of-war over embryonic stem cell research is largely an American phenomenon. Bruce Einhorn, a Hong Kong–based editor for Bloomberg *BusinessWeek*, wrote in his "Eye on Asia" blog that there is "almost zero chance of any such change in policy in Singapore, China, or other Asian countries aspiring to be centers of stem cell research."

FREE EXCHANGE OF STEM CELL KNOW-HOW

It is too soon to know whether human stem cells offer reasonable hope to children and adults with diabetes. It is too soon to know which country will win the stem cell race. What is certain is that the power of biomedical imagination and information technology spills over international boundaries more freely than ever. "Scientific progress is based ultimately on unification rather than fragmentation of knowledge," wrote Fotis C. Kafatos and Thomas Eisner in *Science.* "Kept separate, these domains, no matter how fruitful, cannot hope to deliver on the full promise of modern biology."

The global exchange of stem cell research information is growing in tandem. National and regional networks are already operating in the United States as well as in Australia, Canada, Finland, Germany, Scotland, Switzerland, and Asia-Pacific. And they are funding research. The cutting-edge cord blood stem cell research of Peter Wernet's group at the University of Düsseldorf Medical School in Germany, for instance, has received support from the Stem Cell Network of North Rhine-Westphalia and the stem cell transplant consortium Eurocord, as well as the National Institutes of Health in the United States. Perhaps no regional cooperative is more emblematic of the rise of the life sciences than the ScanBalt BioRegion, a consortium of eleven Scandinavian and Baltic states, northern Germany, and northwest Russia focused on cutting-edge research including regenerative medicine. It is a twenty-first-century expression of the Hanseatic League, an alliance of trading cities in Northern Europe and the Baltic that dominated commerce in the Middle Ages. On the other side of the globe, scientists from Australia, China, India, Japan, Korea, Singapore, Taiwan, and Thailand formed the Stem Cell Network—Asia-Pacific (SNAP) in 2007 with Hong Kong and Indonesia subsequently joining the network.

International standardization and cooperative efforts such as the International Stem Cell Forum and the International Society for Stem Cell Research are invaluable for the evolving science. The International Stem Cell Forum (ISCF—http://www.stem-cell-forum.net) led an early effort to catalog and characterize the world's existing and future embryonic stem cell lines. The International Society for Stem Cell Research (ISSCR—http://www.isscr.org), with more than 3,400 members from fifty-six countries as of 2009, has been at the forefront of developing consensus guidelines for conducting human embryonic stem cell research and its clinical translation, including informed consent documents and material transfer agreement documents. The ISSCR has also taken a leadership role in developing educational programs and addressing the growing problem of stem cell tourism in part through its sponsored website "A Closer Look at Stem Cell Treatments" (http://www.closerlookatstem cells.org).

Stem cell networks spread across the globe ensure a healthy balance of collaboration and competition. So do networked centers of genome sequencing, a field in which scientists are mapping DNA, the source code of all living systems, and a field in which the United States has an overwhelming lead. It is not impossible that, overtime, the amalgamation of brains, research, and competitiveness will reenact some of the "genome wars" of the Human Genome Project in the stem cell field. But "war" in this case is not limited to scientific and economic competition. In the absence of bioweapons treaties, broader global cooperation among nations and their biological scientists may be the world's best safeguard in future wars that will be waged against the body politic by bioterrorists.

Chapter Six

HARBINGERS OF DESTRUCTION

Men out of fear will cling to the thing they most fear.

—Leonardo da Vinci

The Renaissance was a time of discovery, genius, and technological progress. It was also a time of vicious war and conflict. As the noted British art historian Sir Kenneth Clark observed, "During the Renaissance in Italy, the art of warfare was the most important of all the arts, and it marshaled the services of the most skilled artists." Those artists included Giotto, who made drawings of the fortifications of Florence; Michelangelo, who updated Giotto's drawings during a siege in 1529; and the greatest military architect and engineer of them all, Leonardo da Vinci, whose drawings of weaponry include a giant crossbow and a factory for casting giant cannons. It should not be forgotten that the rare genius whose paintings convey the great truths of life and drawings reveal the anatomical truth of the human body also designed catapults, battle tanks, and chariots armed with swirling scythe blades for dissection of a distinctly nonmedical sort.

More than one of Leonardo's military-minded inventions has inspired tools for twenty first-century soldiers. Both the Pentagon's Predator, an aerial drone capable of carrying out surveillance missions and executing precision air strikes, as well as its brain computer interface project, which is designed to allow machines to be controlled through thought power, have roots in Leonardo's genius. The confluence of war and technology continues in the age of the biorenaissance. What will be its ramifications in our modern era of global pandemics and biowarfare? What will be the ultimate outcome when biological tools that can cure turn into biological weapons that can kill? What do stem cells have to do with it?

The impetus for the most massive application of biology in defense and security the world has ever seen was the tragedy of 9/11. The inferno of hatred that erupted that day was sparked decades earlier, on August 29, 1966, when a man in a Cairo prison prayed away the last hours of his life. The prisoner had already served ten years behind bars, having been convicted of plotting against the government. That was a useful charge against fundamentalist opponents of the regime of Gamal Abdel Nasser, and it was widely applied. The sentence was death, a sentence the prisoner welcomed. In his death, he believed that life would be given to his ideas—mainly that the global domination of what he called the "Western system" was coming to an end because "it is deprived of those life-giving values which enabled it to be the leader of mankind." Islam, he believed, was the only system that possesses such values. Only Islam, he was certain, acquainted humanity with a harmonious way of life, whether humanity wanted it or not. Even Islam's golden age of science during the Middle Ages was suspect in the prisoner's reconstruction of history. He disowned Avicenna

(Ibn Sina), known as the "prince of physicians" whose *Qanun* or "Canon of Medicine" formed one of Leonardo da Vinci's scholarly references for his anatomical studies. His counsel to true believers was unambiguous: jihad and the glory of martyrdom.

The prisoner was Sayyid Qutb, founder of the Muslim Brotherhood and the intellectual foundation for Al Quaeda. His hanging on that August day in 1966 set a match to a fuse that burned with increasing fury from the gallows in Cairo to terrorist training camps in Taliban-controlled Afghanistan, to its explosive end at the World Trade Center, the Pentagon, and a field in Pennsylvania. It is a fuse faithfully followed by pediatrician Ayman al-Zawahiri, Osama bin Laden's successor as head of al Qaeda and the best-known living member of the Muslim Brotherhood.

The global war on terror for which Qutb supplied early intellectual and spiritual nourishment was officially declared by President Bush on 9/11 and proceeds relatively unabated. Fears of terrorism in the form of planes have become fears of terrorism in the form of pathogens. And survival in the age of bioterrorism depends on dealing a lethal blow to engineered superpathogens before they can vanquish us. Money was poured into biodefense in the wake of the anthrax attacks in Washington, D.C., that followed 9/11. In the ensuing decade, Congress approved an estimated $50 billion to protect the American people against biological threats from human and natural sources. The life sciences enterprise, both public and private, was funded to deliver vaccines for protection from genetically engineered viruses, to fashion wound-healing aids and brain implant chips for treating injured soldiers, and to design biosensors for detecting and monitoring natural and man-made superpathogens at even the single-molecule level. And that's just for starters.

The fate of all societies against bioterrorism and epidemic onslaughts will hinge on the health of the human immune system. Advances in stem cell technology, as well as biotechnology and nanotechnology, have revealed both the strengths and vulnerabilities of this complex system. Understanding what's at stake not only requires understanding these technologies thoroughly, it also requires that we no longer take our immune systems for granted. Nor can American or indeed Western leadership in science and technology be taken for granted. The era of bioterrorism leaves no room for apathy.

IMMUNITY IN THE AGE OF BIOTERRORISM

Bacteria rule the earth and have done so since their debut several billion years ago. Said biologist Stephen Jay Gould in *The Structure of Evolutionary Theory,* bacteria "were in the beginning, are now, and probably ever shall be (until the sun runs out of fuel) the dominant creatures on earth." If left unchecked, they would take over. As author Michael Crichton observed in his 1969 thriller *The Andromeda Strain,* the mathematics of uncontrolled bacterial growth show that, in a single day, "one cell of *E. coli* could produce a super-colony equal in size and weight to the entire planet Earth." Today, Crichton's claim is used widely as a homework assignment for students studying the mathematics of geometric progression, a subject that intrigued British population growth economist Thomas Malthus two centuries ago.

A healthy immune system has been our body's first and best defense against bacteria since we made our debut as a species on the East African savannah some one hundred thousand years ago. We humans make our worldly debut equipped with maternal antibodies to help us fight infection until our own immune system can get up to speed, at about six months. Without natural immunity, humans would become feedstock for the trillions of bacteria, viruses, funguses, and other microorganisms that reside in or on us. After all, microorganisms make up half of the earth's biomass, more than plants or animals. Most of them typically cause no problems, given the normal functioning immune system, and some—including those that exist in our intestines—even serve a useful purpose. Bacteria are the earth's master chemists. They know what to do with the food we eat.

The human immune system is perhaps the greatest marvel of biological complexity, ranking far ahead of our other physiological systems. Its workings have been described somewhat figuratively as "cognitive"—that is, having an ability to perceive, reason, decide, learn, and remember. In the decades to come, the particulars of how our immune system operates will be captured and re-created in the laboratory. In the digital world, computer scientists and artificial intelligence experts around the world are working to create "artificial" immune systems as a model of an adaptive learning system. Such computer-based models have a host of possible applications, including infectious disease surveillance. That's because healthy immune systems not only have the most prodigious "memories" in nature, but they can also "foresee" the evolution of dangerous microbes and often possess the ability to counter new mutations of microbial threats.

In the past, humankind has been devastated by pandemics of plague, smallpox, tuberculosis, cholera, diphtheria, and other contagious diseases. We like to comfort ourselves with the idea that these are different times. HIV/AIDS, SARS (severe acute respiratory syndrome), H1N1 "swine flu," drug-resistant tuberculosis, and the specter of a bird flu pandemic tell us otherwise. Many experts say it's only a matter of time before we have a global pandemic like bird flu involving a mutated virus that could kill millions, transmitted by infected persons who travel via busy international airports like those in Hong Kong, London, and Chicago. Domestic animals serve as handy mutation chambers for evolving pathogens. "Most major human infectious diseases have animal origins, and we continue to be bombarded by novel animal pathogens," wrote Jared Diamond and colleagues in an article in *Nature* in which they called for a "global early warning system."

Stem cell biology has vastly accelerated our ability not only to follow the development of the immune system as it unfolds in the embryo, but also to create a reasonable replica of the system in a three-dimensional matrix, such as a portable cell-growth cassette or "laptop" system. The "artificial immune system" technology will be an invaluable tool in combating organ transplant rejection and spurring drug development and testing, not to mention developing rapid diagnosis and treatment of emerging infectious diseases.

The future of stem cell science is not only in health care, though that has captured the lion's share of attention. Through their ability to reconstitute immune function in artificial environments, embryonic stem cells, induced pluripotent stem cells, and other stem cell types wield undeniable potential in two fateful arenas: emerging microbial threats and biowarfare. That knowledge is putting humanity in an unprecedented position. Whoever possesses that knowledge possesses a power to destroy that is potentially more pervasive and surely more sinister than its twentieth-century thermonuclear counterpart. Recent advances in using human stem cells to create functioning human immune systems in mice led Pierre Chambon, one of the world's scientific giants, to comment that such research "opens doors into rooms we never thought we could enter."

Before jetliners hijacked by the political descendants of Sayyid Qutb were crashing into the Twin Towers and the Pentagon, the country was still sorting out the Bush administration's new stem cell policy, the topic of his first major speech as president just a month earlier. After 9/11, Bush's speech on stem cells a month earlier was quickly overshadowed and nearly forgotten, except by the scientific community. Yet historians looking back might place equal or greater importance on the subject of his speech, if projections about bioweapons and biodefense are correct.

The concept of defense in the broadest sense, which has weighed heavily on the minds of most Americans since 9/11, serves to "push immunology—the biological system of defense— to the forefront of today's basic research and applied technology," wrote Mike May in *Science*. Among the government agencies that fund applied research in immunology is the one that brought us the Internet, stealth technologies, satellite-enabled global positioning systems, and a variety of other items on the highly classified technological bleeding edge. It's the Department of Defense research arm known as DARPA, and it's the place where some of the country's best minds are brainstorming to come up with ways to harness the tools of the biorenaissance and put them to work for the nation's long-term security. DARPA is eyeing stem cells and living tissues for their potential use in bioprotection through the construction of an artificial human immune system.

MILITARY RESEARCH AND DEVELOPMENT

On October 4, 1957, American leadership was aghast to learn that the Soviet Union had successfully launched *Sputnik*. The radio signals that the satellite sent back to Earth as it orbited overhead served as a reminder of who was on top. In the fit of national anxiety that followed, President Dwight D. Eisenhower and his administration took action by creating the Advanced Research Projects Agency. Established in February 1958, ARPA helped fund Wernher von Braun's Juno rocket program, later renamed Saturn. It was a mighty *Saturn V* rocket that put a man on the moon in July 1969. That both von Braun and Leonardo da Vinci built war machines—in the case of von Braun the German V2 rocket used against London during World War II—accents the fact that "genius for hire" has a long history and it is often brought to bear in conflicts and warfare.

One of ARPA's early luminaries was J. C. R. Licklider, the first head of the agency's computer research program. In the early 1960s, Licklider, a psychologist who had taught at Harvard and then MIT, developed his "galactic network" idea. He envisioned a globally interconnected computer system through which everyone could quickly access data and programs from any site. His ideas and those of his team are reshaping the world we live in. His galactic network—now called the Internet—is the means by which knowledge of discovery is shared and the discovery process is accelerated, in research, innovation, and through informed publics around the globe. It is also increasingly central to the efforts of world health organizations.

Julie Gerberding, director of the Centers for Disease Control and Prevention during the George W. Bush administration, told the National Press Club that "globalization, connectivity, and speed" are the key factors in the way the CDC will deal with emerging global health threats like SARS in the twenty-first century. The changing world described by Gerberding was brought home when the CDC quarantined a U.S. citizen harboring a potent natural bioweapon, drug-resistant tuberculosis, after he ignored a CDC recommendation and flew to Europe to get married. Indeed, in a networked world, the SARS epidemic, the TB quarantine incident, and financial disasters like the Asian currency crash of the late twentieth century are linked together, in the view of Daniel Yergin, author and executive producer of *Commanding Heights*, a TV series about globalization. "A globalized world requires greater cooperation and trust across borders—whether among central bankers or international health authorities," Yergin wrote in "Fighting the Globalization Flu." He might well have added life scientists and counterterrorism experts to his list.

ARPA, renamed DARPA (Defense Advanced Research Projects Agency) in 1972, has funded high-risk and sometimes seemingly remote research with the hope for potentially big payoffs for the military. The fact that much of the research results in failure is "the *expected* price of the quest for unusual breakthroughs," according to a National Academies report. It was this sort of freer system that led to DARPA's development of the Saturn rocket and the M16 rifle (1960s), the Stealth Fighter, global positioning system, and ARPANET/Internet (1980s), the Predator unmanned aircraft (1990s), and the Global Hawk aircraft (2000s). DARPA projects were also influential in developing the National Science Foundation's nanotechnology and computer sciences programs. DARPA has delved into behavioral science, artificial intelligence and speech recognition, high-energy laser technology, organ and limb regeneration via its Restorative Injury Repair program, "suits" with sensors in them that can remotely transmit the severity of a soldier's wound to medical personnel, and a telerobotic surgical machine based on the acclaimed Da Vinci Surgical System. And just as Leonardo drew flight contraptions based on detailed anatomical studies of birds, organisms as diverse as whales, sharks, rodents, bees, butterflies, dragonflies, and moths are being plumbed for useful biological mechanisms beyond the current contribution of animals to camouflage, body armor, detection and decoy strategies, and barb design. DARPA's arsenal of biologically inspired weapons of the twenty-first century feature remote-controlled "cyberinsects."

DARPA picked up a black eye from Congress for foraying into the mysterious new territory of the agency's "Darth Vader" research. In 2003, a DARPA project proposing the development of a future trading market in predictions of assassinations, terrorism, and other events in the Middle East was canceled by then Secretary of Defense Donald Rumsfeld after Congress found out about DARPA's "Total Information Awareness" data-mining program. Its manager, Iran-Contra antihero John Poindexter, was terminated. But what if this harebrained idea had worked and could alert us in some way to better understand risk? What seemingly random acts like terrorism might be predicted? A program like Total Information Awareness might well have allowed the FBI to connect the visa applications of the 9/11 hijackers with their applications to attend flight schools.

A MATTER OF NATIONAL SECURITY

Whatever the logic or illogic behind its schemes, it is DARPA's job to think ahead, far ahead. DARPA's premise is that what others may do can never be known with certainty and so it's best to stay ahead of the game. DARPA got a six-year jump on the mind-set behind the establishment of the Department of Homeland Security when the agency launched its Biological Warfare Defense Program in 1996. The program's goal is to develop technologies to thwart the use of highly resistant biological warfare agents, including bacterial, viral, and bioengineered organisms and toxins, by enemies and terrorists. To achieve it, DARPA wants to find and harness "the fundamental laws of biology" in the words of Brett Giroir when he was director of DARPA's Defense Sciences Office.

In the wake of 9/11, other agencies got into the act. The "National Strategy for Homeland Security," enunciated by President Bush in 2002, stated that the newly created Department of Homeland Security "will unify much of the federal government's efforts to develop and implement scientific and technological countermeasures against human, animal, and plant diseases that could be used as terrorist weapons." Indeed, the Department of Homeland Security itself launched its own version of DARPA, Homeland Security Advanced Research Projects Agency, or HSARPA, and funded it to the tune of hundreds of millions of dollars annually.

But DARPA remains ground zero for visionary thinking. "DARPA is a vehicle to select the most imaginative proposals for research from the academic and industrial communities," said Michael Goldblatt during his tenure as deputy director of the Defense Sciences Office at DARPA in the Clinton administration. But the problem is "too daunting" for democratic large organizations alone. "Small entrepreneurial groups, people with a passion, a tolerance for high-risk, and a commitment to solve a problem, can be much more effective and much more efficient than whole cascades of large organizations," according to Goldblatt.

The coalition of DARPA, HSAPRA, academic scientists, bioengineers, and biotech companies to combat the threat of emerging infectious disease and bioterror is emblematic of the power of biological knowledge in the new century. National defense is, increasingly, biological defense, whether the enemy is a rapidly mutating virus, bacteria, or other pathogen. The even bigger threat lies with those persons around the world trained in the biological sciences who know how to alter the genetic programs of pathogens and convert them into weapons—including weapons of mass destruction—in the era of "garage biology" we are entering, an era when do-it-yourself amateurs as well as laboratory scientists have access both to the powerful tools of molecular and synthetic biology and ready supplies of biological molecules.

"DARPA's strategic thrust in the life sciences, dubbed 'Bio-Revolution,' is a comprehensive effort to harness the insights and power of biology to make U.S. warfighters and their equipment safer, stronger, and more effective," said DARPA director Tony Tether, testifying in March 2003 before the U.S. House of Representatives. "Over the last decade and more, the U.S. has made an enormous investment in the life sciences—so much so that we frequently hear that we are entering a 'golden age' of biology. DARPA is mining these new discoveries for concepts and applications that could enhance U.S. national security in revolutionary ways." Tether's testimony was like an echo from the past, when oil transformed armed conflict in the first half of the twentieth century and electronics brought unprecedented precision to warfare and security in the second half of the century. Now biology was to be yoked to the task of national self-defense. If there were any doubt, in 2006 DARPA launched its "Accelerated Manufacturing of Pharmaceuticals" program to create a rapid manufacturing system for producing vaccines and therapeutic proteins within weeks instead of years. Multimillion dollar contracts were awarded to a consortium of U.S. and British biotechnology and pharmaceutical firms. Program director Michael Callahan, a physician scientist and former State Department bioweapons expert, reminded Congress in 2005 that "biological weapons can be covertly transported as either minute quantities or in a form that leaves no signature, thus allowing the agents to cross international borders and be produced behind enemy lines." The threat of a virulent microorganism is potentially much more ominous than a bombing in a railway station or even planes piloted by suicide "martyrs" flying into office buildings.

The White House itself acknowledged that Americans are entering a dangerous new realm in self-defense in its report "Biodefense for the 21st Century," released in April 2004. Preparing for the new threat is made more difficult by the "dual-use nature of biological technologies," as the report states, and by the potential to conceal these technologies that surpasses the potential to conceal nuclear weapons. In dealing with the potential threat biological weapons pose, the "stakes could not be higher for our Nation." That view was reinforced in a report "World at Risk" issued a month after the election of President Barack Obama in 2008. "To date, the U.S. government has invested the largest portion of its nonproliferation efforts and diplomatic capital in preventing nuclear terrorism," wrote the Commission on the Prevention of Weapons of Mass Destruction Proliferation and Terrorism under the direction of former Senators Bob Graham and Jim Talent. "Only by elevating the priority of preventing bioterrorism will it be possible to substantially improve U.S. and global biosecurity."

Concealment of a virulent bioweapon in an envelope or a parcel is just one method. Antiterror security cannot stop a sophisticated and motivated biotechnology terrorist from exposing international travelers to a fast-spreading virus or a potentially deadly bacterium like XDR (extensively drug resistant) tuberculosis. In more conventional matters, who's inspecting the global supply chain for these types of threats, if anyone at all? What about the estimated seven million containers that enter U.S. ports every year aboard giant cargo ships? In theory, any one them could be carrying a biological payload and a virtual death sentence.

With that growing reality as a backdrop, and recognizing that in its infancy our immune system was a stem cell, DARPA's Defense Sciences Office launched its Engineered Tissue Constructs Program in 2002. The stated goal of the program was "to develop an interactive and functional human immune system" in the laboratory from "a common stem cell source" using tissue-engineering technologies. Such a system would be used to develop and test new vaccines rapidly in a human immune system replica rather than in mice and rats. The program morphed into "Modular Immune In Vitro Constructs," or MIMIC, in 2009, one of three prongs of a new DARPA program called "Blue Angel" designed to speed up the development of vaccines. Writing in *The New Yorker* two years later, David E. Hoffman described Blue Angel as the brainchild of Michael Callahan and his colleagues in response to concern by the White House that a vaccine for H1N1 "swine flu" might not be available in time to prevent a pandemic.

AN IMMUNE SYSTEM IN A BOTTLE

To succeed in creating an artificial human immune system, DARPA and its academic and private sector partners have brought together several critical technologies. Strange as it may seem, commercial inkjet printers have rapidly become a technological staple of biomedical genomics research and bioengineering. They actually print living cells. The printers are capable of depositing cells in precise positions in a three-dimensional matrix of highly specialized bioactive materials that would convert the 3-D environment into a bioreactor. A bioreactor can serve as an artificial organ that mimics, as much as possible, what occurs in the body. Researchers at Carnegie Mellon University have printed a "bio-ink" of stem cells and growth factors that direct their differentiation into specialized cells. Just as Leonardo da Vinci drew muscle and bone during the Renaissance, in the early days of the biorenaissance a "Bio-inkjet Printer Draws Muscle and Bone" ran a story title in *Popular Mechanics*.

In Linda Griffith's MIT laboratory, tissue engineering is being used to create human tissues on silicon chips, like the "liver chip" she developed with $4.4 million in DARPA funding through its Tissue-based Biosensors Program. Created through digital printing, such biochips have the potential to raise the curtain on how disease develops and to accelerate drug development to combat it. Their unique advantage, according to Griffith, is "their responsiveness to unknown agents, such as genetically engineered viruses." Viruses that could kill millions. DARPA is interested in the development of such biochips. The Pentagon and Department of Homeland Security are too.

To help rapidly develop vaccines for bird flu as well as potential bioweapon agents, researchers at MIT, Harvard, and biotech firms funded by DARPA set about in 2002 to mimic the basic operations of the human immune system using digitally printed tissue-engineering systems. Their goal was to develop a bioreactor that can serve not only rapid vaccine assessment needs for biodefense, but also to discern and replicate the instructions the immune system uses to eliminate infectious diseases and pathogens. To build such a system, they are trying to re-create the way molecules and cells assemble themselves in natural systems.

What sets the artificial immune system project apart from other DARPA projects is the new world of biology, including stem cell biology. Stem cells have the potential to transform a thing of science fiction into a plausible reality in the not too distant future. Each stem cell type, such as adult stem cells, embryonic stem cells, and induced pluripotent stem cells, comes with a rapidly growing panoply of reagents that identify them and thereby the biological molecules that promote their growth and differentiation. These reagents serve to reveal changes that stem cells undergo as they morph into more specific cell types. All of which means that the behavior of these powerful tissue-forming cells is increasingly something that can be controlled for the first time in a three-dimensional artificial environment.

The next step for researchers in the field is to attempt to re-create the stem cell microenvironment, the wellspring of regenerative medicine, in bioreactors. And they are doing exactly that. Take chemical engineer J. H. David Wu of the University of Rochester in New York. Wu claimed he has used tissue engineering to mimic the human bone marrow microenvironment, the seedbed of the immune system. His bone marrow bioreactor reportedly can make all types of human immune cells, including natural killer cells and the primitive cells that give rise to them. Just as DARPA was announcing its artificial immune system project, Wu and a colleague filed an application to the U.S. Patent and Trademark Office for a human immune system existing outside the body.

In truth, it is already possible to mimic *some aspects* of human immunity in the laboratory, though not the entire system, at least not yet. We actually have two immune systems to achieve *immunis,* a Latin word meaning "exempt"—to protect us from dangerous pathogens. We have an *innate* immune system and an *adaptive* immune system. Innate immunity is an evolutionary ancient mechanism of host defense found in every multicellular organism, from plants to humans. Innate immunity consists of a set of general responses to disease-bearing pathogens that assault the body. Our skin and the mucus coating our digestive system and airways offer a first line of defense against myriad pathogenic microorganisms. If microorganisms penetrate these barriers, they encounter antimicrobial molecules that restrict the infection.

If the innate immune system does advanced surveillance, the adaptive immune system is where the battle is joined. Its foundation is represented by white blood cells, called *lymphocytes*, which are produced by blood-forming stem cells in the bone marrow. They constitute the two parts of the adaptive immune system in mammals: the *humoral immune system*, which acts by using antibodies in blood and body fluid to attack microorganisms, and the *cellular immune system*, which uses a complex of multiple cell types to ward off external invaders.

From the perspective of a pregnant woman's immune system, an embryo becomes "immunoprivileged," neither rejected as a foreign invader nor attacked by the mother's immune system. Such immunological tolerance may occur because pregnant women experience a unique adjustment in both their innate and adaptive immune systems so that they remain, for the most part, protected from infections and yet can provide protection for the fetus growing within them. The knowledge gained from further study of this maternal-fetal phenomenon could help us learn how to defend ourselves against those who would use against us what normally protects us.

Much of what is known about the human immune system has been learned over the past twenty years. The powerful combination of genomics, bioinformatics, a new technology called RNA interference, and stem cell science will multiply the existing body of knowledge about the human immune system in the coming decades. Thus the idea of a functional artificial immune system for biodefense, which the Defense Department through DARPA is banking on, is not really so far-fetched. Consider the work that has already been accomplished in mice, where the two major components of the immune system—B and T lymphocytes, or white blood cells—have been studied extensively. Now researchers are forging ahead using human cells. They have successfully coaxed human embryonic stem cells into becoming fully functional T cells—the cells that we rely on every day to protect us from the harmful and potentially lethal microorganisms that surround us.

DARPA announced in 2003 that it was funding research on the development of an artificial lymph node in mice. The purpose of the project was to activate the immune system and bolster it against biological agents such as anthrax, smallpox, and natural and genetically engineered bugs, potential bioweapons that don't typically exist in nature. Bioengineers are using tissue-engineered bioscaffolds and seeding the scaffold with what are called dendritic cells, powerful cells of the immune system that are responsible for triggering a T-cell immune response when the body is attacked by invaders. Much like B and T cells, dendritic cells represent one of the most promising avenues for manipulating the immune system—for treating and preventing disease and for giving the body a protective advantage against bioweapons.

This brings us to VaxDesign Corporation, an Orlando, Florida, company and one of DARPA's prime contractors for engineering vaccines and immunotherapies as well as the artificial immune system "for use in protecting American warfighters from Biological Warfare pathogens." Simply put,

VaxDesign is using dendritic cells together with its technology and biomaterials to fabricate artificial lymphoid tissue, what it calls a "vaccine node." It is then using the node to guide the design and engineering of vaccines that activate the adaptive immune system in a precise manner. VaxDesign is in the business of creating fully functional three-dimensional biohybrid tissues for use in an artificial immune system for rapid vaccine assessment.

In 2006, VaxDesign founder and CEO William Warren and members of his scientific team filed patent applications in the United States, Canada, and Europe for an "Automated artificial immune system" to test vaccines, drugs, and biologics. According to the application, "Functional equivalency to the human immune system is achieved by building engineered tissue constructs (ETCs) housed in a modular, immunobioreactor system." In Leonardo-like fashion, Warren is bringing robotics to the task of biodefense. He is using robots to manufacture the surrogate immune systems for vaccines. "What we're trying to do is take a fresh look at immunology by really looking at immunoengineering, applying engineering aspects to immunological challenges," Warren told *First Monday*, a local magazine. Its machines are the centerpiece of DARPA's MIMIC program. The success of VaxDesign's immune system-mimicking technology was borne out when Sanofi Pasteur, the world's largest vaccine manufacturer, purchased the firm in 2010.

DARPA's academic collaborations are also bearing biotech fruit. Chemical engineer Nicholas Kotov at the University of Michigan and his colleagues in a DARPA-funded consortium borrowed from the architectural genius of bees to build an artificial environment to foster the growth of cells and the development of tissue. They used nanotechnology to create a honeycomb scaffold similar to the internal structure of bone marrow that supports stem cells, and researchers looked to see if experimental cells might also thrive there. They found that their 3-D honeycomb scaffold stimulated stem cell interactions. They also were able to control stem cell specialization into the different cell types that make up the immune system, "effectively creating an ex-vivo [outside the body] replica of bone marrow." As news stories reported it, Kotov and his colleagues had created "an immune system in a bottle."

Which brings us back to James Thomson at the University of Wisconsin where human embryonic stem cell lines made their debut in a petri dish in 1998. In 2006, Thomson and his colleagues in a study funded in part by DARPA and published in the *Journal of Immunology* described a method of forming the powerful dendritic cells from human embryonic stem cells. They subsequently filed a patent for the method. The same year in a study also funded by DARPA and published in the journal *Blood*, they reported finding a marker for a unique progenitor cell in the human embryonic stem cell lines Thomson created—cell lines that were later approved by NIH for federal funding. In the development of the embryo, the progenitor cell Thomson's group found gives rise to blood-forming stem cells, the architects of the blood and immune systems. It was one more step along the path to recapitulating human embryonic development in the laboratory, revealing another clue to what only the embryo knows—and what DARPA is determined to find out.

PROJECT BIOSHIELD

After 9/11 and the anthrax attacks in Washington, D.C., it took just three months for the federal government to move on the immunological research front. On December 6, 2001, Health and Human Services Secretary Tommy Thompson announced the launching of seven initiatives designed "to accelerate bioterrorism research and help strengthen the nation's ability to deal with the public health threat posed by bioterrorism." As a harbinger of things to come, the following March the National Institutes of Health announced a funding initiative for "Innovation Grants" proposals that "would have a substantial impact on current thinking and understanding of human immunology," including on infectious disease caused by "agents of bioterrorism." Applicants were encouraged to pursue techniques to study immune cells and how they evolve from progenitor cells. They could use human embryonic stem cell lines approved by the Bush administration the previous August.

The NIH programs created in the aftermath of 9/11 were coordinated by a new biodefense office of the National Institute of Allergy and Infectious Diseases (NIAID), one of the seven institutes that make up the National Institutes of Health. The point man for the new biodefense initiatives has been the federal government's point man for HIV/AIDS, NIAID director Anthony "Tony" Fauci. Fauci's mantra for justifying the massive increase in the NIAID's biodefense budget for 2003—at $1.75 billion nearly eight times that of the year before—was three-tiered. First, the money was needed to bolster translational research and product development to convert basic discovery to biodefense applications. Second, the money was needed to design and develop broader-spectrum therapeutics and vaccines as well as agents targeting specific pathogens. And third, the money was needed to strengthen interactions with the private sector. It would guarantee payment to companies for biodefense products they develop (vaccines, for example), even if some of those products end up stockpiled and had no commercial value under normal circumstances.

These initiatives constituted NIAID's contribution to Project BioShield, announced by President Bush in January 2003, passed by Congress, and signed into law by Bush in July 2004. Project Bioshield provides $5.6 billion over ten years to speed development and stockpiling of vaccines and diagnostics to protect the United States from the worst consequences of a bioterrorist attack. "There is still no defense against the threats that could come from laboratories in the future, such as hybrid diseases, new viruses and bacteria that resist antibiotics," argued *The Washington Post* in an editorial after the bill passed the Senate. Stem cells are one of the technologies at play in vaccine production. In 2007, BioShield awarded a $500 million contract to the Danish firm Bavarian Nordic A/S to produce twenty million doses of its smallpox vaccine for the U.S. Strategic National Stockpile. That firm then obtained a license from the French firm Vivalis to test Vivalis's duck embryonic stem cell lines as a production platform for its smallpox and other vaccines. Vivalis claims its avian stem cells lines could transform traditional egg-based vaccine production. In fall of 2009, as the H1N1 flu was sweeping across the globe, the pharmaceutical giant GlaxoSmithKline partnered with Vivalis to codevelop and manufacture a vaccine for H1N1 using Vivalis's EB66 stem cell line. Following U.S. Food and Drug Administration approval of GSK's application for an investigational new drug (IND) the following year, the British company prepared to launch phase 1 clinical trials to determine the safety of the vaccine, the first to be produced in an avian cell line rather than in chicken eggs, which is the conventional approach to vaccine manufacturing.

Project Bioshield employs the resources of the National Institutes of Health, the Department of Health and Human

Services, and the Food and Drug Administration, including the FDA's Emergency Use Authorization. The purpose of Project BioShield is to expedite research, development, procurement, and availability of biomedical countermeasures against chemical, biological, radiological, and nuclear terrorism. Small biotechnology firms as well as pharmaceutical behemoths began moving aggressively into the biodefense arena well before BioShield was proposed. Perhaps most intriguing of all the BioShield programs is the construction of a system of regional and national "biocontainment laboratories." The laboratories include some Biosafety Level-4 laboratories, facilities that can handle lethal biohazards like the Ebola virus and the Marburg virus. One of the biocontainment laboratories was featured in a story by Mike May published in *Nature Medicine* in 2010: "Stem cells serve as new platform for biodefense preparedness." May described the high-throughput stem cell diagnostics laboratory at Texas A&M University. Researchers are exploring 350,000 mouse embryonic stem cells lines, each with a missing gene, in a search for "genes that offer protection from an onslaught of deadly bacteria, chemicals and viruses."

In May 2007, then Health and Human Services Secretary Michael Levitt brought Project BioShield under the oversight of the newly created Biomedical Advanced Research and Development Authority. BARDA is a program within HHS that "enhances the opportunity for innovation in our efforts to develop effective medical countermeasures against a host of public health threats, either natural or manmade," according to Leavitt. Nearly $500 million was budgeted for BARDA for the fiscal year 2010–2011. Like DARPA, BARDA has an interest in rapid vaccine assessment and development. Unlike DARPA whose "Total Information Awareness" program was outed through a Freedom of Information Act (FOIA) request, BARDA operates completely outside the FOIA. The Health and Human Services secretary or the BARDA director would determine whether a disclosure "would pose no threat to national security," the legislation stated. "Such a determination shall not be subject to judicial review." In other words, the nation's courts have no business tampering in matters of ultimate security and, perhaps, ultimate destiny.

Even as BioShield I experienced a high-profile failure of a vaccine contractor to deliver, behind the scenes in the Senate, BioShield II started to take shape. Senators Joseph Lieberman and Orrin Hatch and others introduced a bill in 2005 that would give immunity from lawsuits for firms creating vaccines or drugs that could cause injuries and also to extend patent protection. "It's easy to imagine how government concerns about patent law could evaporate during a bioterror epidemic," voiced *The Washington Post*. "It's also easy to imagine how that scenario might make companies shy away from the bioterror 'market' in the first place." Yet firms and contractors are lining up. Ground zero for the new bioindustry is Arlington, Virginia, across the Potomac River from Washington, D.C. Arlington is the site of many major conferences that bring together universities, pharmaceutical companies, the biotechnology industry, and the public health community with the national defense establishment. Pulitzer prize–winning *Washington Post* reporter Dana Priest and colleague William Arkin mapped the geography of the post-9/11 national security buildup in an exposé "Top Secret America" in 2010, the result of a two-year investigation. They noted "the underground maze of buildings at Crystal City in Arlington, near the Pentagon" as an example of where defense companies are clustered.

Doesn't President Dwight D. Eisenhower's warning about the dangers of the "military industrial complex" echo on Capitol Hill any more? Has 9/11 muted that presidential caution for good? What does it mean to put biology, the stuff of life itself, into the realm of military competitiveness? What does it mean when biology and stem cells become the stuff not only of health care but of national defense and global geopolitics?

THE FUTURE OF IMMUNITY

NIAID's massive investment in biodefense bears no direct connection with the rapidly emerging world of stem cell research or DARPA's interest in stem cell-mediated artificial immunity. But some in Congress feel that the NIH, as presently structured and managed, is not well suited to lead or even participate effectively in a world of biological weapons and biowarfare.

That was the conclusion of a panel convened for the Strategic Assessments Group by the National Academy of Sciences. That panel wrote that advances in biotechnology, coupled with the difficulty in detecting nefarious biological activity, "have the potential to create a much more dangerous biological warfare (BW) threat." At the core of the threat is the potential lethality of "engineered biological agents" that could be "worse than any disease known to man." The genomic revolution is pushing biotechnology into "an explosive growth phase," a wave of knowledge so complex and moving so rapidly that conventional means for monitoring biological threats are probably already outdated. Indeed, the same NIH disease-focused science "could be used to create the world's most frightening weapons." The panel offered the example of Australian researchers who showed that the virulence of the mousepox virus, a relative of the smallpox virus, can be significantly boosted by engineering into it a gene that interferes with the human immune response. Such a technique could be applied to other pathogens, such as anthrax or smallpox. A novel and lethal virus might even be created from scratch, a virus with no earthly experience, with no history of interaction with the human immune system—a twenty-first-century "Andromeda strain," but one engineered by scientists working for malevolent groups allied against the West. It would be an example of what bioweapons authority Michael Callahan calls next-generation "vaccine-evading biological weapons."

Craig Venter and his colleagues at the Institute for Biological Energy Alternatives, with funding from the Department of Energy, announced in 2004 that they had indeed made a common virus from synthetic DNA and got it to work inside a bacterium. Two years later Venter's team transplanted a genome from one species of bacteria to another and watched it work in its new home, a major step along the pathway to creating a synthetic organism. To the alarm of some, they filed a patent describing "a minimal set of protein-coding genes which provides the information required for replication of a free-living organism in a rich bacterial culture medium." In other words, they sought to patent the genetic minimum for creating life itself. What if this technology were used to create a monster bug for which there is no treatment, no vaccine? One that was highly infectious but took weeks to a month or two to kill people? One such as HIV, which has few signs of an initial infection?

George Church, a Harvard genetics professor and director of the Lipper Center for Computational Genetics, pioneered a direct genomic sequencing method two decades ago and helped get the Human Genome Project off the ground. He is concerned about the potentially deadly combination of the availability of genetic sequence information about dangerous viruses (for example, polio virus and mousepox virus) and the availability of DNA chemicals that can be used to build those sequences, which can be purchased over the Internet. Designing a lethal pathogen, Church believes, has become easier than building a nuclear device, with few controls on the emerging technologies that make building such a pathogen possible. "We want to do for biology what Intel does for electronics," he told *The New York Times*. "We want to design and manufacture complicated biological circuitry."

That includes circuitry for stem cells. "For almost every decision that a stem cell makes, there is an analogous situation in an electronic circuit," said Princeton stem cell scientist Ihor Lemischka. He and colleague Ron Weiss engineered a "toggle switch" system that directed the differentiation of mouse embryonic stem cells to become muscle cells, nerve cells, and insulin-producing pancreatic beta cells. Weiss, now at MIT, is hard at work applying engineering principles to control stem cell decision making in a precise way. More than five thousand cellular and genetic components, modules, circuits, relays, and switches are available to synthetic biologists through an interchangeable biological parts clearinghouse called the BioBricks Foundation involving several top research centers. Foundation funding is provided by, among others, DARPA, the National Science Foundation, and Microsoft Corp.

Today, scientists and engineers are peering under the hood of life, like Leonardo da Vinci delving into the cave, and are rearranging the elements and component parts of the cell, and DARPA is on board. In 2010, the agency invited the synthetic biology community to submit funding proposals through its "Controlling Cellular Machinery—Vaccine Program." Applicants were asked to "describe the development and demonstration of a nucleic acid-based vaccine that uses genetic regulatory elements to control the efficacy of the vaccine." In its drive to disclose the secrets of immunity, DARPA is moving quickly from funding the development of vaccine testing machines to funding the development of next-generation vaccines from cutting-edge science. The problem is that the tools of synthetic biology are increasingly available to amateurs, hobbyists, and do-it-yourself (DIY) biologists.

George Church organized a consortium of researchers and academics to push the federal government to require that anyone interested in purchasing DNA segments that could be used to engineer agents of bioterror be compelled by law to get a license first. "No one without a permit should be doing that research," Church told *Congressional Quarterly* in 2004. That same year the federal government created its own oversight group, the National Science Advisory Board on Biosecurity or NSABB, to provide guidance to federal departments and agencies on the dual-use potential of research advances emanating from the biosciences. When Craig Venter and his research team at the J. Craig Venter Institute reported in 2010 that they had created a synthetic cell using computer code, President Barack Obama called on his bioethics commission to look into the emerging field of synthetic biology and issue a report. After six months of study, the commission recommended that the research proceed with "prudent vigilance" and federal oversight to prevent risk, striking a middle ground between the "precautionary principle"—meaning researchers need to prove that their research is not harmful to the public or the environment—and letting them do whatever research they want without oversight.

Not everyone believes licensing and regulation advocated by Church and others will work. Basic know-how about molecular biology is permeating the educational system, wrote synthetic biologist Rob Carlson in *Biosecurity and Bioterrorism*, and efforts to limit the proliferation of such skills would not only damage the U.S. economy, it would create a black market. Carlson advocates an open system of global networks and professional cross-checking as the best way to ensure security and prevent abuse. It is a call to arms for the research community to police itself. That was what the president's bioethics commission recommended for the do-it-yourself community: "Scrutiny is required to ensure that DIY scientists have an adequate understanding of necessary constraints to protect public safety and security, but at present the Commission sees no need to impose unique limits on this group."

THEORETICAL POSSIBILITIES OF DEATH

The entire foundation of the Manhattan Project in the heyday of nuclear physics nearly seventy years ago rested on little more than a theoretical possibility. The possibility was that splitting the uranium atom would launch a chain reaction releasing energy in vast quantities. In the eyes of the U.S. government, it was a possibility that had to be pursued. No U.S. administration could brook failure to act, certainly not during a world war.

Creating an artificial human immune system is just that, a theoretical possibility. Whoever possesses the power to re-create such a system, to the extent such knowledge is captured, also possesses the presumptive power to circumvent it, to the same extent. Stem cells represent a transit way through the bottleneck separating the world of modern science and technology— the world of genomics and proteomics, systems and synthetic biology, and genetic algorithms, genetically enhanced biomaterials, nanotechnology, and cloning—from the world of engineered human immune system bioreactors. Stem cells, potentially, give us entrée to the world of immune biocomplexity, a world today working perhaps harder than ever before to keep us from merging with natural biosystems at the hands of microorganisms. To understand the immune system sufficiently to re-create it is to possess the potential biological power of annihilation. Such power, such knowledge, has never existed before, not even during the age of thermonuclear weapons. "The bomb was latent in nature as a genome is latent in flesh," wrote atomic bomb historian Richard Rhodes of the days leading up to the Manhattan Project.

Yet there is a distinction to be made between the molecules of elemental matter and the molecules of life. "The release of atomic energy marks a great revolution in human history, but not (unless we blow ourselves to bits and so put an end to history) the final and most searching revolution," wrote Aldous Huxley in a foreword to the 1946 edition of *Brave New World*. "This really revolutionary revolution is to be achieved, not in the external world, but in the souls and flesh of human beings." Knowledge of the sciences of matter—that is, physics, chemistry, and engineering—does not alter the "natural forms and expressions of life itself," Huxley wrote. Knowledge of biology and genetics does. But that was in 1946. Today the "standardization of the human product" under totalitarian management that Huxley anticipated in his great novel remains mercifully distant.

What is not distant is the knowledge of the sciences of matter being brought to bear in biology in an age of bioterror. It is, for example, the engineering of tissue from stem cells so that candidate vaccines can be rapidly tested against bioweapons being developed to evade the immune system. It is not the misdeeds of the "biotechnology project" dreaded by conservatives like bioethicist Leon Kass—the genetic enhancements of Dick and Jane at the hands of their parents and collaborating geneticists— that pose the graver threat. It is the engineering of biology for annihilation and for protection against annihilation by defense agencies and contractors, research establishments, terrorist cells, evildoers and nonstate actors with PhDs. The age-old sword and shield, arrow and armor, bullet and vest, and bomb and shelter will be reenacted in a *danse macabre,* this time with bioweapons and vaccines.

Just as science is a global enterprise, so is stem cell research, which is being undertaken at universities and research institutes throughout the United States. But it is also flourishing at the Scottish Centre for Regenerative Medicine in Edinburgh and the Centre for Stem Cell Biology at the University of Sheffield in Great Britain, at the RIKEN Center for Developmental Biology in Kobe and the Nara Institute near Hiroshima in Japan, at Biopolis in Singapore, at Seoul National University in South Korea, at the Technion in Israel, at the Karolinska Institute and the Universities of Göteborg and Lund in Sweden, and at McMaster University in Ontario, Canada. It is flourishing as far north as Biocenter Oulu in Finland above the Arctic Circle near the border with Lapland, and as far south as Monash University in Australia. Stem cell research is also well under way in China, India, and Russia. These three countries alone account for 2.6 billion people, or about 40 percent of the total world population. The West has the technological lead for the moment, but these countries have the brainpower, the will, and the bodies. China and Russia have a legacy of bioweapons research. And all three are home to nonstate actors with 9/11 motivation, if not yet the know-how, to put biology in the service of their glorious cause. In an ironic twist of history given the Soviet Union's aggressive bioweapons program, then Russian President Vladimir Putin banned export of all medical specimens including blood, skin, and organs from his country in 2007. According to news reports, Putin was worried that the genetic information of his countrymen could enable Western scientists to make ethnic-specific biological weapons designed to target Russians including agents that would cause infertility among Russian women.

In 2003, in a bioweapons report prepared for the Central Intelligence Agency, a life sciences expert panel warned of the dangers of dual-use technologies. It emphasized that it will be extremely difficult to distinguish between legitimate biological research activities and the experimental production of potential advanced bioweapon agents. That is in contrast to detecting nuclear weapons. A nuclear weapon "has always had clear surveillance and detection 'observables,' such as highly enriched uranium or telltale production equipment," such as centrifuges and other separation devices. Most of the panelists believed that a different relationship between the government and life sciences communities might be needed to grapple effectively with the future bioweapons threat. It is the only way to deal with "the pace, breadth, and volume of the evolving bioscience knowledge base, coupled with its dual-use nature and the fact that most is publicly available via electronic means and very hard to track."

The ability to make bioweapons is well within the reach of individuals who have an educational foundation in molecular biology, microbiology, biochemistry, immunology, and a wide variety of other biological disciplines. More people with such a foundation exist today than have ever existed. Even more will exist tomorrow, products of a global and ever-expanding higher-education enterprise.

Indeed, the construction of sophisticated biocontainment facilities could actually increase the likelihood of a terrorist attack involving the use of biologic weapons, in the view of Louise Richardson, a leading scholar on international terrorism and vice chancellor of the University of St. Andrews in Scotland, which has a strong research center for the study of terrorism and political violence. At present, the key scarcity among terrorists is "a dearth of adherents with the skills to understand and deploy biologic weapons," wrote Richardson in the *New England Journal of Medicine*. "The operation of these facilities will require the training of scores, if not hundreds, of people to work with deadly pathogens. By training more experts in biologic weapons, we are increasing the probability that one or more of them will have sympathies with a terrorist organization." The world got a sneak preview of this just after 9/11 in the anthrax attacks in Washington, D.C. The anthrax was not garden-variety; there is a strong possibility based on circumstantial evidence that the anthrax was prepared by U.S. Army bioweapons scientist Bruce Ivins in his laboratory at the Army's Medical Research Institute of Infectious Diseases at Fort Detrick, Maryland. Ivins committed suicide in 2008.

According to the authors of one study, biotechnology techniques and equipment readily available on the open market make the large-scale production of bioweapons in small-scale facilities quite feasible and at relatively low cost. Meanwhile, the knowledge of how our immune system works grows exponentially, thanks to new fields like stem cell biology. At the same time, the tools for devising ingenious ways to circumvent it are becoming available to an even larger number of people with potentially hostile intent.

They've proven they can be quick learners. Making a lethal bio-weapon may be more complex than learning to pilot an airliner for a short flight, but there's no reason to hurry. As Bill Gates once said, "We always overestimate the change that will occur in the next two years and underestimate the change that will occur in the next ten."

CONTROLLING THE POWER OF STEM CELLS

We live in an age when applied scientific knowledge has produced what the late Stephen Jay Gould called the "great asymmetry." Moments can undo what only centuries can build—cities, gothic cathedrals, rain forests, ancient statues of Buddha carved into a mountain on the Silk Road in Afghanistan blown to bits by Taliban antiaircraft guns and dynamite. Gould cited familiar agents of wartime destruction, from gunpowder to nuclear bombs and the impact of technology on global biosystems. Yet human choice, not the "intrinsic content of science," determines whether science becomes a potent agent of benevolent change or "an accelerator of destruction on the wrong side of the great asymmetry." Therefore, scientists have a special responsibility to provide counsel "rooted in expertise," just as military strategists have a responsibility to take into account asymmetric warfare.

Scientists have the responsibility, but do they have the audience? "Suppressing research into stem cells is causing that research to move abroad, which will damage America's biotechnology industry," wrote *The Economist* in "Cheating Nature." "But that will not be fatal to America's future, and opponents of stem-cell research might argue that it is a price worth paying for their beliefs. Monkeying with defense is a different matter. . . .

America's current military prowess has been achieved, in large part, because the country has listened to and lauded its physicists and engineers." Will America listen to its biological scientists and bioengineers?

It is too soon to know when advances in the biological sciences will bring about a comparable sense of urgency in the realm of governance as they begin to emerge in the realm of defense and public safety. But it will happen, and probably sooner rather than later. Nation-states initially will deal with the rise of new knowledge as they did when the atom was revealed to contain secrets of vast energy release and weapons of mass destruction. Then after due deliberation about how to sequester these immense stores of new knowledge for national defense and commercial advantage, they will experience a moment of clarity. They will be reminded once again that even the most sophisticated technology springing from that knowledge is an itinerant creature, a bedouin, potentially a threat in the wrong hands. Keeping it safely at home has always been a dubious proposition.

Countries with advanced science and technology will invest to stay ahead in the bioweapons knowledge game, just as the human immune system itself, in normal times, stays ahead of the mutating organisms that would destroy it. Biotechnology and biodefense companies will benefit from national investment. But overtime, a bioweapons arms race would become autoimmunity at work on the global body politic, civilization turning on itself in a deadly corporeal competition of power-seeking.

Is a stem cell–based artificial immune system possible? One muggy Sunday in July 1939, physicists Leo Szilard and Eugene Wigner paid a visit to Albert Einstein at Little Peconic Bay on Long Island, where he was spending the summer. They broke the news to Einstein about the recent experimental work with neutrons. They suggested it would be possible to use a neutron, discovered only seven years earlier, to start a chain reaction in the uranium nucleus. His response? "I never thought of that." The man who launched a revolution in physics had not imagined that his famous equation of 1905, $E=mc^2$, might be used to release massive amounts of energy in a weapon of war. Six years later to the day, July 16, 1945, the nuclear chain reaction that the world's greatest scientific mind had never considered was initiated at the "Trinity" atomic bomb test site on a desert flat northwest of Alamogordo, New Mexico.

Physicists called the chain reaction "fission" after the process called "binary fission" in which one bacterium divides in two. As evolutionary biologist Richard Dawkins has observed, heredity itself began as a lucky self-regenerating chain reaction. Thus the nomenclature of nuclear disintegration was named for the reproductive process of the living matter that "rule the earth." Bacteria. In an experiment that may portend the future of data storage, Japanese researchers converted the letters "$E=mc^2$ 1905!" into genetic sequences and inserted them into the genome of *Bacillus subtilis*, a bacterium with four million DNA letters. The code of special relativity had entered the code of life. Fission spread the word.

What would it mean to capture the power of the human immune system, a system that took evolution millennia to create? The drive to disclose the secrets of the stem cell is relentless. Just as war accelerated the harnessing of the power of the atom, the prospects of pandemic flu as well as bioterror accelerate the application of molecular and stem cell biology for bioprotection. Who will provide guidance for the wise use of such power if it is placed in our hands? Is there someone of the stature and wisdom of Niels Bohr to guide us through the dangerous, uncharted territory revealed by our collective genius? Bohr was the Danish physicist and Nobel laureate who urged Franklin Roosevelt and Winston Churchill to commit to the peaceful development of nuclear power, but failed.

Bohr was a philosophical descendant of the ancient Roman poet Lucretius who wrote that bodies are composed of minute, indivisible particles of matter called atoms. Lucretius wrote in his poem *On the Nature of Things* during the darkest days of Rome: "The dread and darkness of the mind cannot be dispelled by the sunbeams, the shining shafts of day, but only by an understanding of the outward form and inner workings of nature." No one knew that better than Leonardo da Vinci. If you wish to know the inner workings of the human body, he wrote, "you—or your eye—require to see it from different aspects, considering it from below and from above and from its sides, turning it about and seeking the origin of each member; and in this way the natural anatomy is sufficient for your comprehension."

Among scientists, the search for knowledge is paradoxically an article of faith. That faith that the universe is knowable is at the heart of the scientific quest. How else, asks immunologist and author Gerald Callahan, can you explain the unshakable belief of scientists "that we will one day understand something as infinitely complicated as our own immune systems when we don't yet understand the character of the atom, or our own confidence that their tinkering with the parts of something so powerful and so intricate will inevitably lead to greater benefit and understanding?"

As surely as stem cells brought us to where we are individually and collectively as human beings, our future and our fate is inextricably linked to what happens in stem cell research laboratories around the world. That reality is just beginning to sink in. What we do about it—what we do to ensure that the fruits of the research are directed to benevolent and nonideological ends—is up to us.

BEYOND THE DARKNESS

Why does the eye see a thing more clearly in dreams than with the imagination being awake?
—Leonardo da Vinci

The conflicting human emotions of hope and fear that Leonardo da Vinci felt at the entrance of that cave are always amplified during times of technological change. Hope and fear are positioned at opposite ends of a social tension line, a line under stress from forces wanting to push forward and forces wanting to hold back. The tension is constant and viscerally felt, especially in advanced technological societies.

Leonardo himself eventually entered the cave through what his biographer Charles Nicholl calls his "flights of the mind." The cave is the setting for some of his most famous paintings like the Louvre's *The Virgin of the Rocks*, which art historians say displays the kinship he felt with the natural world. The scene in the sandstone grotto is a "geological tour de force" in the words of one critic, an expression of Leonardo's acute visual grasp of the anatomy of the earth. His experience in the cave would launch him on an intellectual journey, wrote Yale University surgeon and author Sherwin Nuland. It would convince Leonardo that the earth and its living forms were much older than the biblical account of creation and are constantly changing.

One day Leonardo converted what had danced in the human imagination for eons—human flight—into a doodle. The drawing was a direct and intentional violation of what was then known about gravity, about human beings and their place in the natural order. Sketched around 1478, when Leonardo was in his twenties, the drawing depicts a flying machine with reticulated wings that could be guided by a suspended pilot. The doodle was, in essence, that of a hang glider.

Nothing captures the spirit of the Renaissance better than the spirit of flight. It was a time of flights into the artistic and scientific unknown, flights often forbidden by authorities of the spiritual and temporal realms of the time. Flights human beings have been reluctant to relinquish ever since. Today, such flights of the mind are responsible for air travel, productive agriculture, central heating, public health, global wealth, and voyages to the moon, Jupiter, and Saturn. They are responsible for the boom in cell phone use in sub-Saharan Africa, secure electronic communications in papal conclaves, televangelism and sports TV worldwide, satellite-enabled global positioning systems, and computer-enabled genome mapping. Flights of the mind are responsible for bringing the scourge of polio under control, creating test tube babies like Louise Brown, and giving people with a wide range of diseases hope for a better life, including a healthy kid named Molly Nash.

On a summer day in 1978, the summer Louise Brown was born and five hundred years after Leonardo drew his doodle, twenty-nine-year-old Marilyn Hamilton prepared to launch herself off a cliff half a mile above the valley of Tollhouse Mountain near Fresno, California. She had done it countless times before, experiencing the exhilaration of flight over and over again. But this time she forgot to secure herself in the hang glider. As it lifted, she slipped. She grabbed the control bar to keep from falling off. That act of desperation put her glider into a nosedive. She crashed, breaking her back. The accident left her unable to walk.

Some stories like Marilyn Hamilton's end there, with permanent disability the unfortunate price of trying to fly like a bird. The medieval norm in the West was to accept one's fate and atone for the sins of presumption and despair. But Hamilton was not about to settle for that. She and two hang glider companions built a line of lightweight, high-performance wheelchairs that folded up flat, influenced by aircraft wing design. The innovators sold their highly successful business to Fresno-based Sunrise Medical in 1986, the year after Hamilton was named California's businessperson of the year. She also became a U.S. wheelchair tennis champion and wheelchair skiing athlete. In her own way, Hamilton followed in the footsteps of Leonardo by her impatient inventiveness. She is often quoted as saying, "If you can't stand up, stand out."

One day in April 2005, some two hundred wheelchairs stood out on the Capitol Hill lawn in Washington, D.C. They carried people of all ages from all over the United States. They suffered from spinal cord injury. They were there to support passage of the Christopher Reeve Paralysis Act, legislation that would provide funding to establish a clinical trial network for persons with spinal cord injury. Though the proposed legislation did not include stem cell research, Reeve's name is inextricably linked to the field. Leading the charge was Dana Reeve, chair of her late husband's foundation and a cancer victim who soon would lose her own battle. She was determined to put legs of research advocacy under his legacy. "Although Chris has died, I know the work he did has not died," she told the crowd. His wheelchair, in Reeve's view, was something to get out of, not something to get used to.

Marilyn Hamilton and Christopher Reeve embraced Leonardo's legacy of human individuality and courage. Both took a chance and entered the dark cave of the unknown, a cave Leonardo revisited in a later and graver endeavor, to see if the spirit and the biomaterials of humans might be joined in the body, to see how the eye was "the window on the soul" and how the brain was "the seat of the soul," as he put it.

Leonardo often considered the soul in the context of nature. He viewed trees, which he painted with exquisite detail, as feeling creatures, endowed with spirit. Indeed, the tree of life is a ubiquitous cultural symbol, its branches reaching skyward while its roots dig deep into the earth. It bears seeds that renew the life cycle, seeds of bounty, seeds that themselves hold the seeds of perpetual renewal, of a seemingly endless sequence. At the beginning of the biorenaissance the question is inescapable: is the tree of life being sustained or exploited by stem cell research, cloning, gene sequencing, genetic engineering, and other biological technologies?

Today, scientists are hard at work sequencing the complete tree of human life going back into prehistory. It is a task described by James Watson in *Time* magazine cover story published in 2003 commemorating the fiftieth anniversary of Watson's codiscovery with Francis Crick of the structure of DNA. With each passing year, more is being revealed about how a single embryonic stem cell at the trunk of the tree of life makes branching decisions, deciding what kind of cell to be, what kind of tissue or organ to make, what kind of limb to build, and how to apply the finishing touches of skin, hair, and nails. That information will spur the new field of regenerative medicine. But at what price? What do we make of a world in which "science has not only scaled the Tree of Life but chopped it down and begun trading its fruit as a commodity?" asks Todd Aglialoro, the editor of the Catholic Sophia Institute Press. Are we deforesting our future as a species? Or are we pursuing an essential path both to extending life and health and ensuring our own survival?

In the twenty-first century, global power will be reflected more and more in the power of biological technologies, whoever possesses them. These technologies arise and move like online music and video file-sharing. "The advent of the home molecular biology laboratory is not far off," wrote research scientist Rob Carlson, a keen observer of trends and projects in applied and synthetic biology, the type of projects often funded by DARPA. Moving genes between organisms "might be slightly more complicated than baking cookies, but it is for the most part less complicated than making wine or beer," Carlson said, noting that discussions about the potential applications of cellular cloning "are barely underway." Carlson's colleague Oliver Morton, writing in *The New York Times,* cautioned that biologists should start talking to one another, "before the trees of knowledge in their synthetic garden bear their strange fruit," so they know what to do "when the serpent turns up." The bioethical marathon has just begun. All political, social, and religious institutions need to take that reality into account.

A few months before Cardinal Ratzinger became Pope Benedict XVI, he called the power of cloning "more dangerous than weapons of mass destruction." Inflammatory sentiments, but he wanted to make a point about human tampering with the sacred. Yet nature consigns most fertilized eggs to oblivion by ejecting them from women rather than implanting them in a uterine wall to begin a pregnancy. Does that fact not give human beings license to intervene in the natural scheme of re-production or biological discharge? Such interventions, including in vitro fertilization, cannot be condoned in the Vatican's eyes. "Man is capable of producing another man in the laboratory who, therefore, is no longer a gift of God or of nature," Ratzinger said. "He can be fabricated and, just as he can be fabricated, he can be destroyed."

There is indeed potentially destructive power in stem cell research and cloning. There is an immense threat to people already walking the Earth. It resides in the power of human eggs and applied human biology to create and creatively destroy—to fabricate the materials of human beings that can be used in medical treatments or as knowledge platforms in a biodefense race against pathogens created at home like computer viruses, knowledge platforms like artificial human immune systems.

In the biorenaissance, we are already fabricating components of the human body in the laboratory using stem cell technology, cellular cloning, inkjet printing, and other bioengineering techniques. We can justifiably imagine re-creating organ components and the systems that they serve—the immune system, the nervous system, the circulatory system, and others. That knowledge can serve those with intent for good or evil. All nations engaged in research using biological technologies know that. The current debate is about new medical treatments and possibly cures from stem cell research versus the concern that nascent human life is being treated as a commodity in the process. The coming debate will be about whether to move first in a world of nation states and terrorist cells possessing bioagendas. The argument will be about what to do, and when, to secure nations, cultures, ideologies, and faiths. A united world community may influence whether and how the power of the new biology is deployed. But that power is not likely to be curtailed by condemnations from the Vatican, the Southern Baptist Convention, or the U.S. Department of State, not to mention nonbinding United Nations declarations. The noble passion that drove Marilyn Hamilton and Christopher Reeve to scale the heights of human possibility has a deadly reciprocal side, as we saw on 9/11.

In the early years of the biorenaissance, we are poised in front of the dark cave peering into it, shifting right and left to gain the advantage of the ever-changing light. We hear voices of hope, desperate for news from the tree of knowledge. We hear voices of fear, convinced we are about to founder in a turbulent ethical sea. It is tempting to think that the biorenaissance will be guided by another Leonardo, someone with the profound gift to know whether to enter the dark caves we will encounter on our journey. Will he or she be an American or a Brit, born of Western culture and values? Or will he or she be a Chinese, a descendant of the ancient Han people, a child of the Middle Kingdom? Or will he or she be an Indian attuned to the Hindu cycle of rebirth and regeneration, or possibly a Singaporean who works in the gleaming science city Biopolis? Or, more likely, will that leadership come from a team of people collaborating around the world, not from one individual or laboratory?

What is certain is that the biorenaissance will unfold increasingly in one world. It is a world we are creating through communications technologies, international organizations, international travel, international finance, and global environmental interdependency. Our world is becoming one world through our response to the threat from global pandemics and terrorism, and through the earth-flattening effects of global economic competition. Most of all, we are in one world through the biology and the Earth we share. People must understand the power of the life sciences and act collectively to shed light on its darker side.

Can facing life's challenges—a child with a devastating genetic disease or a parent with Alzheimer's—be an acceptable reason for reexamining the boundaries of our moral space? Will hope push us forward into the cave, or will fear hold us back? Will the quest to discover "the marvelous thing within" be worth the risk, or will we find that the darkness was better left alone? Or is something else at play, as suggested by playwright Mary Zimmerman in "The Notebooks of Leonardo da Vinci." Does the womblike cave convey a deep longing to return to the place of our origin—to the mystery Leonardo portrayed in *The Fetus in the Womb*?

Whether to challenge our beliefs and assumptions about what it possible, what we consider ethical, and what is wise is ours to decide. We determine how hard to push back the frontiers of knowledge in pursuit of a longer and healthier life for ourselves, our children, and our grandchildren. Day by day we are unveiling the mystery of the stem cell, one of the uniquely generative and regenerative forces in the universe. The biorenaissance gives us the power to bring creative action to the realms of discovery, health, longevity, even human destiny, in ways scarcely imagined. Are we, the inheritors of Leonardo's relentless curiosity about how nature works, prepared to understand that power, seize it, and use it wisely?

GLOSSARY*

adult stem cell: a versatile nonspecialized cell found in a differentiated tissue like skin. Adult stem cells can renew themselves and can (with certain limitations) differentiate to yield all the specialized cell types of the tissue from which it originated.

algorithm: a rule or set of rules for solving a problem.

allele: variant forms of the same gene. Different alleles produce variations in inherited characteristics such as eye color or blood type.

allogeneic transplant: transplanted tissue that comes from someone other than the intended recipient of the transplant. Typically, the donor is a sibling or someone whose tissue type is a close match to that of the transplant recipient.

Alzheimer's disease: a degenerative brain disease of unknown cause that is the most common form of dementia.

amino acid: any of the twenty-six chemical building blocks of proteins.

amyloid plaque: a waxy translucent substance consisting of protein. A buildup of these plaques is believed to interfere with the transmission of neural messages in the brain and lead to the death of brain cells involved in memory, as in Alzheimer's disease.

amyotrophic lateral sclerosis (ALS): a rare degenerative disease that affects motor neurons, leading to progressive muscular weakness and death. ALS is also called Lou Gehrig's disease.

antibody: an infection-fighting protein molecule in blood or body fluids that tags, neutralizes, and helps destroy pathogenic microorganisms (e.g., bacteria, viruses) or toxins.

antigen: any substance that stimulates the immune system to produce antibodies.

assisted reproductive technologies (ARTs): fertility treatments or procedures that involve laboratory handling of gametes (eggs and sperm) or embryos. An example of ARTs is in vitro fertilization.

autoimmune disease or disorder: a category of diseases and disorders in which one's own cells are mistakenly identified as foreign by the body and are therefore attacked by the immune system, causing tissue damage. Multiple sclerosis is an autoimmune disease.

autologous transplant: transplanted tissue that comes from the intended recipient of the transplant. Such a transplant helps avoid the complications of immune rejection.

B lymphocyte (B cell): one of the two major classes of lymphocytes, B lymphocytes are white blood cells of the immune system made by the bone marrow and spleen. B cells develop into plasma cells, which produce antibodies.

bacterium: a single-celled organism. Bacteria are found throughout nature and can be beneficial or cause disease.

bioengineering: the application of engineering principles to the fields of biology and medicine. Bioengineers develop devices that substitute for defective or missing body tissues and organs. Also called biomedical engineering.

bioethics: the study of the ethical and moral implications of new biology, as in the fields of genetic engineering and stem cell research.

bioinformatics: the merger of biotechnology and information technology with the goal of revealing new insights and principles in biology and genetics.

biotechnology: the use of living organisms or their products including DNA, cells, and enzymes to make products or solve problems.

blastema: a mass of nonspecialized or undifferentiated cells from which an organ or body part develops.

blastocyst: a preembryo in mammals, consisting of one hundred to two hundred cells, that has not implanted in a uterus.

bone marrow: the soft, living tissue that fills most bone cavities and contains hematopoietic (blood-forming) stem cells from which arise all red and white blood cells in the body.

bone marrow transplantation: a medical procedure to replenish the soft tissue within bones that produces new blood cells. Bone marrow transplants are necessary when the bone marrow has been destroyed by drug or radiation therapy for cancer, often leukemia.

cancer: diseases in which abnormal cells divide and grow unchecked. Cancer can spread from its original site to other parts of the body and can also be fatal if not treated adequately.

cell culture: growth of cells in the laboratory for experimental research.

cell division: method by which a single cell divides to create two cells. This continuous process allows a population of cells to increase in number or maintain its numbers.

cell-based therapies: treatment in which stem cells are induced to differentiate into the specific cell type required to repair damaged tissues.

cell line: a defined population of cells that has been maintained in a laboratory culture for an extended period and keeps dividing (replicating).

cell lineage: a term used to describe cells with a common ancestry. Cells that developed from the same stem cell have the same lineage.

cell-mediated immunity (cellular immunity): the immune response coordinated by T cells. This branch of the immune system targets cells infected with microorganisms such as viruses, fungi, and certain bacteria.

cerebellum: the portion of the brain in the back of the head between the cerebrum and the brain stem. The cerebellum controls balance for walking and standing, and other complex motor functions.

chromosome: one of the threadlike "packages" of genes and other DNA in the nucleus of a cell. A normal human body cell (somatic cell) has forty-six chromosomes; a normal human reproductive cell (gamete) has twenty-three chromosomes.

clone: an exact replica. A clone may be an exact genetic copy or replica of a DNA molecule, cell, tissue, organ, or entire plant or animal. Dolly the sheep was a genetic clone of her mother.

cord blood stem cells: see *umbilical cord blood stem cells.*

dedifferentiation: the process in which a differentiated cell loses its special form or function or reverts to an earlier developmental stage.

dendritic cell: a specialized type of immune cell that plays a crucial role in recruiting T cells to fight off invasion by a foreign body.

Deoxyribonucleic acid (DNA): the material inside the nucleus of cells that carries genetic information.

diabetes mellitus: a disease in which the body does not properly control the amount of sugar in the blood.

differentiation: the process in which an unspecialized early embryonic cell or adult stem cell becomes a specialized cell, such as a heart, liver, or muscle cell following cell division.

directed differentiation: manipulating stem cell culture conditions to cause a stem cell to divide and become a specialized cell, such as a heart, liver, or muscle cell.

DNA replication: the process by which the DNA double helix unwinds and makes an exact copy of itself.

DNA sequencing: the scientific technique for determining the exact order of the base pairs in a segment of DNA.

double helix: the structural arrangement of DNA, which looks something like an immensely long ladder or two pieces of ribbon twisted into a helix, or spiral coil.

ectoderm: upper, outermost layer of a group of cells derived from the inner cell mass of the blastocyst or preembryo. It gives rise to skin, nerves, and the brain.

egg: the mature female reproductive cell.

embryo: in humans, the developing organism from the time of fertilization until the end of the eighth week of gestation when it is called a fetus.

embryoid bodies: rounded collections of cells that can arise when embryonic stem cells are cultured in the laboratory. Embryoid bodies contain cell types derived from all thee germ layers or fundamental tissues of the embryo: the endoderm, the mesoderm, and the ectoderm.

embryonic germ (EG) cells: pluripotent stem cells that are derived from early germ cells (those that would become sperm and eggs). EG cells are thought to have properties similar to embryonic stem cells.

embryonic stem (ES) cells: nonspecialized cells from the embryo that have the potential to become a wide variety of specialized cell types (that is, are pluripotent). They come from the inner cell mass of the blastocyst.

embryonic stem (ES) cell lines: embryonic stem cells that have been grown in the laboratory under conditions that allow them to self-renew without becoming specialized cells.

endoderm: lower layer of a group of cells derived from the inner cell mass of the blastocyst. The endoderm gives rise to lungs and various organs in the abdomen such as the liver and pancreas.

enzyme: a protein that encourages a biochemical reaction, usually speeding it up. Organisms could not function if they had no enzymes.

Fanconi anemia: a rare inherited disorder in which the bone marrow does not make blood cells. Symptoms include frequent infections, easy bleeding, and extreme fatigue. People with Fanconi anemia have an increased risk of developing certain types of cancer.

fertilization: the process whereby male and female gametes (sperm and egg) unite.

fetus: a developing human from approximately eight weeks after conception until the time of its birth.

gamete: an egg (in the female) or sperm (in the male) cell.

gastrulation: an early stage of development in which large groups of cells move in a coordinated fashion to give rise to the ectodermal, mesodermal, and endodermal germ layers.

gene: a functional unit of heredity that is a segment of DNA in a specific site on a chromosome. A gene directs the formation of RNA, which in turn directs the formation of a protein.

genetic engineering: the laboratory technique of recombining genes to produce proteins used for drugs and vaccines.

genome: the complete genetic material of an organism.

germ cell or germline cell: a sperm or egg, or a cell that can develop into a sperm or egg. All other body cells are called *somatic cells.*

germ layers: the three initial tissue layers arising in the embryo—endoderm, mesoderm, and ectoderm—from which all other body tissues develop.

gestation: the period of development of an organism from fertilization of the egg until birth.

growth factors: special proteins that serve as signaling molecules for controlling cell growth and differentiation or specialization.

hematopoietic stem cell: a stem cell from which all red blood cells, white blood cells, and platelets develop.

Human Genome Project: an international research project to map each human gene and to completely sequence human DNA.

immunity: natural or acquired resistance provided by the immune system to a specific disease.

implantation: the process by which an embryo becomes attached to the inside of the uterus after fertilization. This usually happens after seven to fourteen days in humans.

in vitro: Latin: literally, "in glass," in a laboratory dish or test tube, in an artificial environment.

in vitro fertilization (IVF): a technique that unites the egg and sperm in a laboratory, instead of inside the female body.

induced pluripotent stem cells (iPS cells): adult cells such as skin cells that are genetically reprogrammed to behave like embryonic stem cells in the laboratory. Like embryonic stem cells, iPS cells can make all three germ layers of tissue: the endoderm, the mesoderm, and the ectoderm.

informed consent: a process by which a patient gives written consent (agreement) to undergo a medical procedure after having been provided by his or her doctor with information about the nature of the procedure and its risks, potential benefits, and alternatives.

inner cell mass: the cluster of cells inside the blastocyst that include embryonic stem cells.

islet cells: cells located in pancreatic islets or islets of Langerhans of the pancreas that produce insulin. The cell type within the islets that does this is called a beta cell.

IVF: in vitro fertilization (qv).

leukemia: cancer of the developing blood cells in the bone marrow.

lymphocyte: a type of white blood cell that is part of the body's cellular immune system. It is found in the blood and lymphatic tissue.

Mendel, Johann (Gregor): Austrian biologist who laid the foundations for the science of genetics.

mesenchymal stem cells: multipotent adult stem cells that can differentiate into a variety of cell types including muscle, cartilage, fat, and bone.

mesoderm: middle layer of a group of cells derived from the inner cell mass of the blastocyst; it gives rise to bone, muscle, and connective tissue.

microenvironment: the molecules and compounds, such as nutrients and growth factors, that surround a cell and are important in determining the characteristics of the cell.

mitochondria: small bodies in the cell that provide energy to the cell.

molecular biology: the study of the structure, function, and makeup of biologically important molecules.

molecule: the smallest particle of a substance that retains all the properties of the substance and is composed of a number of atoms.

multiple myeloma: a cancer of the bone marrow formed of one of the bone marrow cells such as white blood cells or plasma cells.

multipotent stem cells: stem cells that can give rise to a number of different cell types. Compare *pluripotent stem cells.*

mutation: a change in DNA that alters a gene and thus the proteins the gene makes. Mutations can lead to disease. They can occur spontaneously during cell division or can be triggered by environmental stresses, such as sunlight, radiation, and chemicals or through heredity.

myelin: the fatty substance that covers and protects nerves.

neural stem cell: a stem cell found in adult nerve tissue that can give rise to the different types of cells of the nervous system.

neurons: nerve cells; the structural and functional unit of the nervous system.

nuclear transfer: a procedure in which a nucleus from a donor cell is transferred into an egg cell from which the nucleus has been removed. See *somatic cell nuclear transfer.*

nucleus: the compartment of a cell that contains the chromosomes, the *X*-shaped gene-bearing bodies.

Parkinson's disease: a progressive neurological disorder that results from the death of nerve cells in a region of the brain that controls movement.

placenta: an organlike structure that develops in the womb or uterus during pregnancy. The placenta serves to anchor the embryo or fetus after implantation.

plasticity: the ability of stem cells from one adult tissue to generate the differentiated cell types of another tissue.

pluripotent stem cells: stem cells that include in their progeny all cell types that can be found in a postimplantation embryo, fetus, or developed organism. Compare *multipotent stem cells.*

polymerase chain reaction (PCR): a technique for making multiple copies of a specific stretch of genetic material. PCR can be used to test for the presence of genes or mutations in DNA.

preimplantation genetic diagnosis (PGD): before an in vitro–fertilized embryo is implanted in a woman's womb or uterus, it can be screened for specific genetic mutations that are known to cause severe genetic diseases. One or more cells are removed from the preimplantation embryo for testing, which causes no harm to the embryo.

primitive streak: a structure that forms during the early stages of embryonic development. It is considered the forerunner of the nervous system.

primordial germ cell: a cell that makes eggs or sperm.

proliferation: expansion of a population of cells by the continuous division of single cells into two identical daughter cells.

protein: a large complex molecule made up of one or more chains of amino acids. Proteins perform a wide variety of activities in the body.

red blood cell: a cell that carries oxygen to all parts of the body.

regeneration: the regrowth of a lost tissue or body part.

regenerative medicine: a treatment in which stem cells are induced to differentiate within or outside the body into the specific cell types required to repair damaged or depleted adult cell populations or tissues.

reprogramming: resetting the developmental clock of a nucleus of an adult cell so that it can carry out the genetic program of an early embryonic cell nucleus—making the proteins required for embryonic cell development.

ribonucleic acid (RNA): a chemical that is similar in structure to DNA. One of its main functions is to translate the genetic code of DNA into structural proteins.

signals: internal and external factors that control changes in cell structure and function.

somatic cell: any cell of a plant or animal other than a reproductive cell.

somatic cell nuclear transfer (SCNT): a procedure in which a nucleus from a donor cell is transferred into an egg cell from which the nucleus have been removed. SCNT is the scientific term for cloning.

sperm: mature male reproductive cells.

stem cell bank: a storage facility for stem cell lines.

stem cells: nonspecialized cells that have the dual capacity to divide indefinitely (self-renew) and also to differentiate into more mature cells with specialized functions.

substantia nigra: a region of the brain that produces dopamine, a chemical that regulates voluntary movement. The loss of dopamine-producing cells leads to the muscular rigidity and tremors characteristic of Parkinson's disease.

telomerase: the enzyme that directs the replication or rebuilding of telomeres.

telomere: the specialized structure at the end of a chromosome that is involved in the replication and stability of the chromosome.

teratoma: a benign tumor made up of embryonic *germ layers.*

tissue-specific stem cell: an adult stem cell found in a specific tissue, like heart tissue, and committed to making new cells for that tissue.

transcription factor: any of various proteins that bind to DNA and play a role in the regulation of gene expression by promoting transcription.

transdifferentiation: the observation that stem cells from one tissue may be able to differentiate into cells of another tissue.

translation: The process of turning instructions from messenger RNA (mRNA), base by base, into chains of amino acids that then fold into proteins. This process takes place in the cytoplasm, on structures called ribosomes.

transformation: a genetic process resulting in a heritable alteration of the properties of a cell. In the case of cultured cells, transformation often refers to the acquisition of new properties, such as unlimited culture life span.

umbilical cord blood stem cells: stem cells collected from the umbilical cord at birth that can produce all of the blood cells in the body (hematopoietic).

undifferentiated: not having developed into a specialized cell or tissue type. A stem cell is an undifferentiated cell.

uterus: the womb; a muscular pear-shaped organ in which the fetus develops.

vaccine: a preparation that stimulates an immune response that can prevent an infection or create resistance to an infection.

vascular: composed of or having to do with blood vessels.

virus: an infectious microorganism composed of a piece of genetic material—RNA or DNA—surrounded by a protein coat.

white blood cells: blood cells that do not contain hemoglobin. White blood cells include lymphocytes, neutrophils, eosinophils, macrophages, and mast cells. These cells are made by bone marrow and help the body fight infection and cancer.

*Compiled from the National Academies reports "Stem Cells and the Future of Regenerative Medicine" (2002) and "Scientific and Medical Aspects of Human Reproductive Cloning" (2002), the glossaries from the National Institutes of Health, the American Association for the Advancement of Science, the Oak Ridge National Laboratory, and other resources. Some definitions have been simplified by the authors.

TIMELINE*

1859

Principle of natural selection propounded: Charles Darwin's *On the Origin of Species by Means of Natural Selection, or the Preservation of Favored Races in the Struggle for Life* is published.

1865

Heredity shown to be transmitted in units: Austrian monk and botanist Gregor Mendel's experiments on peas demonstrate that heredity is transmitted in discrete units. The understanding that genes remain distinct entities even if the characteristics of parents appear to blend in their children explains how natural selection could work and provides support for Darwin's theory.

1869

DNA isolated: Swiss physician Frederick Miescher isolates DNA from cells for the first time and calls it "nuclein."

1879

Mitosis described: German scientist Walther Flemming describes chromosome behavior during animal cell division. He stains chromosomes to observe them clearly through a microscope and describes the whole process of mitosis in 1882.

1895

The term "stem cell" is coined: German zoologist Valentin Häcker coins the term "stem cell" (*Stammzelle*) to designate the cell in the early embryo of the crustacean *Cyclops* that gives rise to the primordial germ cells—the stem cells of the germ line eggs and sperm used in reproduction.

1900

Rediscovery of Mendel's work: Botanists deVries, Correns, and von Tschermak-Seysenegg independently rediscover Mendel's work while doing their own work on the laws of inheritance. The increased understanding of cells and chromosomes at this time allowed the placement of Mendel's abstract ideas into a physical context.

1909

The word "gene" is coined: Danish botanist and geneticist Wilhelm Johannsen coins the word "gene" to describe the Mendelian unit of heredity. He also uses the terms "genotype" and "phenotype" to differentiate between the genetic traits of an individual and its outward appearance.

1911

Chromosomes found to carry genes: Thomas Hunt Morgan and his students study fruit fly chromosomes. They show that chromosomes carry genes, and also discover genetic linkage.

1938

Cloning experiments published: German embryologist Hans Spemann publishes the results of his primitive nuclear transfer (cloning) experiments involving salamander embryos.

1944

DNA shown to transforms cells: Oswald Avery, Colin Mac-Leod, and Maclyn McCarty show that DNA (not proteins) can change the properties of cells, clarifying the chemical nature of genes.

1952

First animal cloning: American developmental biologists Robert Briggs and Thomas J. King clone northern leopard frogs.

1953

DNA structure revealed: British molecular biologist Francis H. Crick and American molecular biologist James D. Watson, working at Cambridge University, describe the double helix structure of DNA. They receive the Nobel Prize for their work in 1962.

1962

Reprogramming of gene expression by nuclear transfer: British biologist John Gurdon extends the research of Briggs and King by cloning a frog using nuclei from the somatic cells of a tadpole. His experiment showed for the first time that genes are not lost or changed during cell differentiation but just differentially expressed.

1966

Genetic code cracked: American biochemist and geneticist Marshall Nirenberg and others figure out the genetic code that allows nucleic acids with their four-letter alphabet to determine the order of twenty kinds of amino acids in proteins.

1968

First successful bone marrow transplant: University of Minnesota physicians perform the world's first successful bone marrow transplant. Stem cells in the donor's marrow reconstruct the patient's blood-forming system.

1970

Embryonic stem cells found: Leroy Stevens at the Jackson Laboratory in Bar Harbor, Maine, proposes the existence of pluripotent embryonic stem cells after observing strange cells in mouse embryos that formed teratomas, growths made up of various tissues such as bone, skin, and teeth.

1973

First recombinant DNA organism: American biochemists Stanley Cohen and Herbert Boyer create the first recombinant DNA organism using recombinant DNA techniques pioneered by Paul Berg. The "gene splicing" technique allows scientists to manipulate the DNA of an organism.

1975

DNA sequencing: Two groups—British biochemist Frederick Sanger and colleagues, and American molecular biologists Alan Maxam and Walter Gilbert—both develop rapid DNA sequencing methods. In the Sanger method, colored dyes are used to identify each of the four nucleic acids that make up DNA.

1976

First genetic engineering company: Biochemist Herbert Boyer and venture capitalist Robert Swanson found Genentech. The California-based company produces the first human protein in a bacterium, and by 1982 markets the first recombinant DNA drug: human insulin.

1978

First baby born through in vitro fertilization: Louise Joy Brown, the first baby to result from in vitro fertilization, is born in Oldham, England, the culmination of years of research by Cambridge University physiologist Robert G. Edwards and Oldham gynecologist Patrick C. Steptoe. Edwards was awarded the Nobel Prize in 2010.

1981

Pluripotent stem cell lines from mouse embryos: Martin Evans and Matthew Kaufman at the University of Cambridge in England and Gail Martin at the University of California in San Francisco identify, isolate, and culture pluripotent embryonic stem cells from mouse embryos.

1983

PCR invented: The polymerase chain reaction, or PCR, is used to amplify DNA. This method, invented by scientist Kerry Mullis while he was working at Cetus Corporation, allows researchers to quickly make numerous copies of a specific segment of DNA, enabling them to study it more easily.

1987

First human genetic map produced: The first comprehensive genetic map is based on variations in DNA sequence that can be observed by digesting DNA with restriction enzymes. Such a map can be used to help locate genes responsible for diseases.

1988

Hematopoietic (blood-forming) stem cell identified in humans: Pathologist and developmental biologist Irving Weissman of Stanford University identifies the human hematopoietic (blood-forming) stem cell, opening the way for research and treatment of many different kinds of cancers and blood diseases.

1990

The Human Genome Project launched: The Department of Energy and the National Institutes of Health announce a plan for a fifteen-year project to sequence the human genome. This will eventually result in sequencing all 3.2 billion letters (chemical subunits of DNA) of the human genome.

1997

First mammal cloned: Dolly the sheep, the first mammal cloned from an adult cell, is produced in Scotland. Ian Wilmut and colleagues report the birth of a lamb derived from the transfer of a nucleus obtained from an adult ewe udder cell into an enucleated oocyte (egg) that was then implanted into a surrogate mother.

1998

Embryonic stem cells isolated from human blastocysts: Wisconsin biologist James Thomson, writing in the journal *Science*, reports the first ever isolation and culturing of human embryonic stem cells derived from human embryos after in vitro fertilization.

2001

Publication of working draft of the human genome: Special issues of *Science* and *Nature* contain the working draft of the human genome sequence. *Nature* papers include initial analysis of the descriptions of the sequence generated by the publicly sponsored Human Genome Project under the direction of Francis Collins, while *Science* publications focus on the draft sequence reported by the private company Celera Genomics under the direction of J. Craig Venter.

2001

U.S. embryonic stem cell policy established: President George W. Bush announces a policy that will permit scientists to use federal funds to study embryonic stem cell lines derived prior to the president's announcement of August 9, 2001.

2001

Directed differentiation of human embryonic stem cells: As human embryonic stem cell lines are shared and new lines are derived, more research groups report methods to direct the differentiation of the cells in vitro. Many of the methods are aimed at generating human tissues for transplantation purposes, including pancreatic islet cells to treat diabetes, neurons that release dopamine to treat Parkinson's disease and cardiac muscle cells to treat heart attacks or heart failure.

2002

Stem cell–based artificial immune system: In the wake of 9/11 and the anthrax attacks in the Washington, D.C., area, the

Defense Advanced Research Projects Agency (DARPA) announces a research program "to develop the technologies and science for supporting efforts leading to the creation of a three-dimensional ex vivo [laboratory-based] human immune system" using human stem cells. The system will be designed to test experimental vaccines that could counter novel biowarfare agents.

2003

Completion of human genome sequencing: Completion of the human genome sequence is announced on the fiftieth anniversary of Watson and Crick's published description of the DNA double helix. The finished human genome sequence is estimated to be at least 99.99 percent accurate.

2004

First cloned human embryonic stem cells reported: South Korean scientists Hwang Woo-suk and Shin Yong Moon report evidence of a pluripotent human embryonic stem cell line derived from a cloned blastocyst. Later the research was found to be faked.

2004

California Institute for Regenerative Medicine (CIRM) established: California voters pass Proposition 71 that establishes a constitutional right to conduct stem cell research while prohibiting funding of reproductive cloning. The $3 billion bonding initiative authorizes creation of the California Institute for Regenerative Medicine to make the state a global leader in stem cell research.

2005

Human neural stem cells repair spinal cords in mice: California scientists successfully regenerate damaged spinal cord tissue and improve mobility in mice with damaged spinal cords using adult human neural stem cells.

2005

World's first cloned dog is born: The world's first cloned dog, an Afghan named "Snuppy," is born. Snuppy was created by South Korean scientist Hwang Woo-suk and his research team. Following an investigation of fraud concerning Dr. Hwang's research, in 2006 Korean scientists confirm that Snuppy is a true clone.

2006

U.S. Congress passes the Stem Cell Research Enhancement Act: The U.S. Congress passes the Stem Cell Research Enhancement Act of 2005 that would significantly increase federal support for embryonic stem cell research. President Bush vetoes the bill.

2007

The California Institute for Regenerative Medicine begins to distribute grants for stem cell research: After more than two years of legal delays, the California Institute for Regenerative Medicine begins to distribute research grants following a California Supreme Court ruling that agreed with a lower court decision upholding the constitutionality of Proposition 71.

2007

Japanese and American scientists create pluripotent stem cells in mice without destroying embryos: Japanese and American scientists reprogram mouse skin cells back to the embryonic state, bringing the field of regenerative medicine a step closer to its ultimate goal of converting a patient's cells into specialized tissues that could replace those lost to disease.

2007

U.S. Congress passes the Stem Cell Research Enhancement Act: The U.S. Congress passes the Stem Cell Research Enhancement Act of 2007. President Bush again vetoes the bill. At the same time he issues an executive order requiring federal agencies to pursue alternative avenues of research on pluripotent cells that do not involve the destruction of embryos.

2007

The Nobel Prize for Physiology or Medicine is awarded to scientists working with embryonic stem cells: The 2007 Nobel Prize for Physiology or Medicine is awarded to American scientists Mario Capecchi and Oliver Smithies and the British scientist Sir Martin J. Evans "for their discoveries of principles for introducing specific gene modifications in mice by the use of embryonic stem cells."

2007

A primate is cloned for the first time: Reproductive biologist Shoukhrat Mitalipov and his research team at the Oregon Primate Research Center report that they have cloned embryos from a rhesus macaque monkey and derived two embryonic stem cell lines from the embryos.

2007

Human skin cells are reprogrammed into pluripotent stem cells: Physician and geneticist Shinya Yamanaka in Japan and geneticist Junying Yu and developmental biologist James Thomson in Wisconsin report their research teams have successfully reprogrammed human skin cells to become pluripotent stem cells. These "induced pluripotent stem cells," or iPS cells, are shown to create all three germ layers found in embryonic development: endoderm, mesoderm, and ectoderm.

2008

Recreation of animal heart with stem cells: University of Minnesota biomedical scientist Doris Taylor and her research team recreate a rat heart using a process called perfusion decellularization, which removes all cells from the heart. The scaffold left behind is seeded with stem cells and progenitor cells in a bioreactor and the heart begins to beat with aid of a pacemaker.

2009

U.S. embryonic stem cell policy revised: President Barack Obama issues an executive order removing federal funding limitations placed on human embryonic stem cell research by the Bush administration and orders the National Institutes of Health to develop new guidelines and safeguards for such research.

2009

World's first human clinical trial of embryonic stem cell–based therapy is launched: Geron Corporation announces that the U.S. Food and Drug Administration (FDA) has granted clearance of the company's Investigational New Drug (IND) application for the clinical trial of its embryonic stem cell–based therapy for patients with acute spinal cord injury.

2010

Court injunction halts U.S. funding of embryonic stem cell research: U.S. federal district court judge Royce Lamberth, citing the 1996 Dickey-Wicker Amendment that bans federal funding of research involving the creation or destruction of human embryos, issues an injunction against federal funding for embryonic stem cell research.

2011

Federal appeals court panel overturns lower court injunction: The U.S. Court of Appeals for the District of Columbia Circuit reverses Judge Lamberth's decision, allowing federal funding of human embryonic stem cell research to continue.

*Compiled largely from the National Institutes of Health, the National Academy of Sciences, the Institute of Medicine, and other public sources.

BIBLIOGRAPHY

PROLOGUE: INTO THE CAVE

Green, Miranda Jane. *Celtic Myths*. Austin: British Museum Press/ University of Texas Press, 1993.

Leonardo daVinci. *Anatomical Drawings from the Royal Library, Windsor Castle*. New York: Metropolitan Museum of Art, 1983.

———. *The DaVinci Notebooks*. New York: Arcade Publishing, 2005.

McGlone, J. "It's incredibly moving to have the piece of paper that Leonardo drew and wrote upon himself." *Scotland on Sunday,* November 24, 2004.

Schultheiss, D., et al. "The Weimar anatomical sheet of Leonardo daVinci (1452–1519): An illustration of the genitourinary tract." *BJU International* 84 (1999): 595–600.

Toledo-Pereyra, L. H. "Leonardo daVinci: The hidden father of modern anatomy." *Journal of Investigative Surgery* 15 (2002): 247–49.

CHAPTER 1: AGENTS OF HOPE

Baker, M. "Stem cell therapy or snake oil?" *Nature Biotechnology* 23 (2005): 1,467–69.

Bissonnette, C. J., et al. "The controlled generation of functional basal forebrain cholinergic neurons from human embryonic stem cells." *Stem Cells.* Published online March 4, 2011: DOI: 10.1002/stem.626.

Bliss, M. *The Discovery of Insulin.* University of Chicago Press, 1982.

Cataldi, L., and F. Vassilios. "Leonardo daVinci and his studies on the human fetus and the placenta." *Acta Biomed Ateneo Parmense* 71 (Suppl. 1, 2000): 405–6.

Center for International Blood and Marrow Transplant Research. *CIBMTR Report of Survival Statistics for Blood and Marrow Transplants.* CIBMTR, 2006.

Chase, M. "Diamond Minds: Research shows how brain power helps athletes excel." *Wall Street Journal* (classroom edition), January 2003.

Chen, M. R., et al. "Generation of pluripotent stem cells from patients with type 1 diabetes." *Proceedings of the National Academy of Sciences U.S.A.* 106 (2009): 15,768–73.

Committee on the Biological and Biomedical Applications of Stem Cell Research, National Research Council, and Institute of Medicine. *Stem Cells and the Future of Regenerative Medicine.* Washington, D.C.: National Academy Press, 2002.

Deshpande, D., et al. "Recovery from paralysis in adult rats using embryonic stem cells." *Annals of Neurology* 60 (2006): 22–34.

Diamond, Jared. "The double puzzle of diabetes." *Nature* 423 (2003): 599–602.

Dimos, J. T., et al. "Induced pluripotent stem cells generated from patients with ALS can be differentiated into motor neurons." *Science* 321 (2008): 1,218–21.

The Earth Institute. *A Race against Time: The Challenge of Cardiovascular Disease in Developing Economies.* New York: Columbia University, 2004.

Greenbaum, Linda E. "From skin cells to hepatocytes: advances in application of iPS cell technology." *Journal of Clinical Investigation* 120 (2010): 3,102–5.

Groopman, Jerome. "Annals of Medicine: The Reeve Effect." *New Yorker,* November 10 (2003): 82–93.

Hawking, Stephen. *A Brief History of Time.* New York: Bantam Books, 1988.

Herbert, L. E., et al. "Alzheimer disease in the U.S. population: Prevalence estimates using the 2000 census." *Archives of Neurology* 60 (2003): 119–22.

Hewson, M. A. *Giles of Rome and the Medieval Theory of Conception: A Study of the "De formatione corporis humani in utero."* London: Athlone Press, 1975.

Jackson-Leach, R., and T. Lobstein. "Estimated burden of paediatric obesity and co-morbidities in Europe. Part 1.The increase in the prevalence of child obesity in Europe is itself increasing." *International Journal of Pediatric Obesity* 1(2006): 26–32.

Kahn, J. "Making lives to save lives." CNN Interactive —Health: Ethics Matters, posted October 16, 2000. http://edition.cnn.com/2000/HEALTH/10/16/ethics.matters/index.html (accessed November 2, 2007).

Kaufman, D., et al. "Hematopoietic colony-forming cells derived from human embryonic stem cells." *Proceedings of the National Academy of Sciences U.S.A.* 98 (2001): 10,716–21.

Kelleher, R. J., et al." Translational control by MAPK signaling in long-term synaptic plasticity and memory." *Cell* 116 (2004): 467–79.

Kemp, Martin. "Vincian Velcro." *Nature* 396 (1998): 25.

Kiatpongsan, S., and D. Sipp. "Monitoring and regulating offshore stem cell clinics." *Science* 323 (2009): 1,564–5.

Kögler, G., et al. "A new human somatic stem cell from placental cord blood with intrinsic pluripotent differentiation potential." *Journal of Experimental Medicine* 200 (2004): 123–35.

Leonardo daVinci. *The DaVinci Notebooks.* New York: Arcade Publishing, 2005.

Matsui, W. H., et al. "Characterization of clonogenic multiple myeloma cells." *Blood* 103 (2004): 2,332–36.

National Institute of Neurological Disorders and Stroke. NINDS Parkinson's Disease. http://www.ninds.nih.gov/disorders/parkinsons_disease/parkinsons_disease.htm (accessed November 2, 2007).

Nistor, G. I., et al. "Human embryonic stem cells differentiate into oligodendrocytes in high purity and myelinate after spinal cord transplantation." *Glia* 49 (2005): 385–96.

Ott, H. C., et al. "Perfusion-decellularized matrix: using nature's platform to engineer a bioartificial heart." *Nature Medicine.* Published online January 13, 2008 [DOI:10.1038/nm1684].

Patel, A. N., et al. "Surgical treatment for congestive heart failure with autologous adult stem cell transplantation: A prospective randomized study." *Journal of Thoracic and Cardiovascular Surgery* 130 (2005): 1,631–38.

Perin, E. C., et al. "Improved exercise capacity and ischemia 6 and 12 months after transendocardial injection of autologous bone marrow mononuclear cells for ischemic cardiomyopathy." *Circulation* 110 (2004): 213–18.

Perrier, A. L., et al. "Derivation of midbrain dopamine neurons from human embryonic stem cells." *Proceedings of the National Academy of Sciences U.S.A.* 101 (2004): 12,543–48.

Rashid, S. T., et al. "Modeling inherited metabolic disorders of the liver using human induced pluripotent stem cells." *Journal of Clinical Investigation* 120 (2010): 3,127–36.

Schmidt, D., et al. "Prenatally fabricated autologous human living heart valves based on amniotic fluid derived progenitor cells as single cell source." *Circulation* 116 (2007): 164–70.

Segev, H., et al. "Differentiation of human embryonic stem cells into insulin-producing clusters." *Stem Cells* 22 (2004): 265–74.

Shen, Q., et al. "Endothelial cells stimulate self-renewal and expand neurogenesis of neural stem cells." *Science* 304 (2004): 1,338–40.

Shriver, Maria. *What's Happening to Grandpa?* New York: Little, Brown, 2004.

Slack, J. M. W. *Stem Cells—A Very Short Introduction.* Oxford University Press, 2012.

Song, H. J., C. F. Stevens, and F. H. Gage. "Neural stem cells from adult hippocampus develop essential properties of functional CNS neurons." *Nature Neuroscience* 5 (2002): 438–45.

Spangrude, G. J., S. Heimfeld, and I. L. Weissman. "Purification and characterization of mouse hematopoietic stem cells." *Science* 241 (1988): 58–62.

Steinbrook, R. "The Cord-blood-bank controversies." *New England Journal of Medicine* 351 (2004): 2,255–57.

Strauer, B. E., et al. "Repair of infarcted myocardium by autologous intracoronary mononuclear bone marrow cell transplantation in humans." *Circulation* 106 (2002): 1,913–18.

Takahashi, K., and S. Yamanaka. "Induction of pluripotent stem cells from mouse embryonic and adult fibroblast cultures by defined factors." *Cell* 126 (2006): 663–76.

Temple, S. "The development of neural stem cells." *Nature* 414 (2001): 112–17.

Thomson, J. A., et al. "embryonic stem cell lines derived from human blastocysts." *Science* 282 (1998): 1,145–47.

Voltarelli, J. C., et al. "Autologous nonmyeloablative hematopoietic stem cell transplantation in newly diagnosed type 1 diabetes mellitus." *Journal of the American Medical Association* 297 (2007): 1,568–76.

Wild, S., et al. "Global prevalence of diabetes." *Diabetes Care* 27 (2004): 1,047–53.

Willerson, J. T. "The future of heart health." *CV Network* (International Academy of Cardiovascular Sciences), August 2004.

Wolf, E. F., et al. "Endometrial stem cell transplantation restores dopamine production in a parkinson's disease model." *Journal of Cellular and Molecular Medicine.* Published online April 7, 2010 [DOI: 10.1111/j.1582-4934.2010.01068.x].

World Health Organization. The World Health Report 2004: Changing History. http://www.who.int/whr/2004/en/ (accessed November 2, 2007).

Yan, Y., et al. "Directed differentiation of dopaminergic neuronal subtypes from human embryonic stem cells." *Stem Cells* 23 (2005): 781–90.

Young, Y., et al. "Clonally expanded novel multipotent stem cells from human bone marrow regenerate myocardium after myocardial infarction." *Journal of Clinical Investigation* 115 (2005): 326–38.

Zohlnhofer, D., et al. "Stem cell mobilization by granulocyte colony-stimulating factor in patients with acute myocardial infarction: A randomized controlled trial." *Journal of the American Medical Association* 295 (2006): 1,003–10.

CHAPTER 2: ARCHITECTS OF DEVELOPMENT

Adams, D. S., A. Masi, and M. Levin. "H+ pump-dependent changes in membrane voltage are an early mechanism necessary and sufficient to induce tail regeneration." *Development* 134 (2007): 1,323–35.

Agata, Kiyokazu. "Utilizing knowledge gained from planarians in regenerative medicine." *RIKEN News* 270 (2003): 12.

Alexander, B. "The morph man." *Los Angeles Times Magazine,* April 4, 2004.

Allen, G. E. "Essays on science and society: Is a new eugenics afoot?" *Science* 294 (2001): 56–61.

———. *Thomas Hunt Morgan: The Man and His Science.* Princeton, N.J.: Princeton University Press, 1978.

Brawley, C., and E. Matunis. "Regeneration of male germline stem cells by spermatogonial dedifferentiation in vivo." *Science* 304 (2004): 1,331–34.

Bryne, J. A., et al. "Producing primate embryonic stem cells by somatic cell nuclear transfer." *Nature* 450 (2007): 497–502.

Carrel, Alexis. "On the permanent life of tissues outside of the organism." *Journal of Experimental Medicine* 15 (1912): 516–28.

————, and Charles Lindbergh. *The Culture of Organs.* New York: Paul B. Hober, 1938.

Chambers, I., et al. "Functional expression cloning of *nanog*, a pluripotency sustaining factor in embryonic stem cells." *Cell* 113 (2003): 643–55.

Dalley, S. *Myths from Mesopotamia.* Oxford: Oxford University Press, 1989.

Darwin, Charles. *The Voyage of the Beagle: Journal of Researches into the Natural History and Geology of the Countries Visited During the Voyage of H.M.S. Beagle Round the World.* New York: Modern Library, 2001.

Da Silva, S. M., P. B. Gates, and J. P. Brockes. "The newt ortholog of CD59 is implicated in proximodistal identity during amphibian limb regeneration." *Developmental Cell* 3 (2002): 547–55.

Dinsmore, C. E., ed. *A History of Regeneration Research: Milestones in the Evolution of a Science.* Cambridge, England: Cambridge University Press, 1991.

Egli, D., et al. "Developmental reprogramming after chromosome transfer into mitotic mouse zygotes." *Nature* 447 (2007): 679–85.

Finkel, E. "Researchers derive stem cells from monkeys." *Science NOW Daily News,* June 19, 2007.

Friedman, David M. *The Immortalists: Charles Lindbergh, Dr. Alexis Carrel, and Their Daring Quest to Live Forever.* New York: Ecco, 2007.

Gerami-Naini, B., et al. "Trophoblast differentiation in embryoid bodies derived from human embryonic stem cells." *Endocrinology* 145 (2004): 1,517–24.

Gould, Stephen Jay. "What only the embryo knows." *New York Times,* August 27, 2001.

Greider, C. W., and E. H. Blackburn. "A telomeric sequence in the RNA of *Tetrahymena* telomerase required for telomere repeat synthesis." *Nature* 337 (1989): 331–37.

Hayflick, L. "Mortality and immortality at the cellular level: A review." *Biochemistry* (Mosc) 62 (1997): 1,180–90.

Hodges, A. *Alan Turing: The Enigma.* New York: Simon & Schuster, 1983.

Horder, T. J., J. A. Witkowski, and C. C. Wylie, eds. *A History of Embryology*. Cambridge, England: Cambridge University Press, 1985.

Hummer, A., et al. "Liver regeneration after adult living donor and deceased donor split-liver transplants." *Liver Transplantation* 10 (2004): 374–78.

Hyun, I., et al. "New advances in iPS cell research do not obviate the need for human embryonic stem cells." *Cell Stem Cell* 1, no. 4 (2007): 367–68.

Ito, M., et al. "Wnt-dependent de novo hair follicle regeneration in adult mouse skin after wounding." *Nature* 447 (2007): 316–20.

Johnson, Paul. *The Renaissance: A Short History*. New York: Modern Library, 2000.

Khosrotehrani, K., et al. "Transfer of fetal cells with multilineage potential to maternal tissue." *Journal of the American Medical Association* 292 (2004): 75–80.

Kielman, M. F., et al. "Apc modulates embryonic stem cell differentiation by controlling the dosage of ß-catenin signaling." *Nature Genetics* 32 (2002): 594–605.

Kinzler, K. W., et al. "Identification of FAP locus genes from chromosome 5q21." *Science* 253 (1991): 661–65.

Kragl, M., et al. "Cells keep a memory of their tissue origin during axolotl limb regeneration." *Nature* 460 (2009): 60–65.

Lagasse, E., et al. "Purified hematopoietic stem cells can differentiate to hepatocytes in vivo." *Nature Medicine* 11 (2000): 1,229–34.

Lewis, R. "A stem cell legacy: Leroy Stevens." *Scientist* 14 (2000): 19.

Li, W., et al. "Generation of rat and human induced pluripotent stem cells by combining genetic reprogramming and chemical inhibitors." *Cell Stem Cell* 4 (2009): 16–19.

Luigi, W., et al. "Highly efficient reprogramming to pluripotency and directed differentiation of human cells using synthetic modified mRNA." *Cell Stem Cell* 7 (2010): 618–30.

Maherali, N., et al. "Directly reprogrammed fibroblasts show global epigenetic remodeling and widespread tissue contribution." *Cell Stem Cell* 1:1 (2007): 55–70.

Mitsui, K., et al. "The homeoprotein nanog is required for maintenance of pluripotency in mouse epiblast and ES cells." *Cell* 113 (2003): 631–642.

Morgan, T. H. *Regeneration.* Columbia University Biological Series. New York: Macmillan, 1901.

Nusse, Roel, and Harold E. Varmus. "Many tumors induced by the mouse mammary tumor virus contain a provirus integrated in the same region of the host genome." *Cell* 31 (1982): 99–109.

Odelberg, Shannon, et al. "Dedifferentiation of mammalian myotubes induced by msx1." *Cell* 103 (2000): 1,099–1,109.

Pearson, Helen. "The regeneration gap." *Nature* 414 (2001): 388–90.

Polan, Mary Lake, and Mylene W. M. Yao. "Stem cell transfer: The egg teaches the chicken." *Journal of the American Medical Association* 292 (2004): 104–5.

Reya, Tannishtha, et al. "A role for Wnt signalling in self-renewal of haematopoietic stem cells." *Nature* 423 (2003): 409–14.

Rodda, S. J., et al. "Embryonic stem cell differentiation and the analysis of mammalian development." *International Journal of Developmental Biology* 46 (2002): 449–58.

Rosenthal, N. "Prometheus's vulture and the stem cell promise." *New England Journal of Medicine* 349 (2003): 267–73.

Shamblott, M. J., et al. "Derivation of pluripotent stem cells from cultured human primordial germ cells." *Proceedings of the National Academy of Sciences U.S.A.* 95 (1998): 13,726–31.

Shen, S., et al. "Dedifferentiation of lineage-committed cells by a small molecule." *Journal of the American Chemical Society* 126 (2004): 410–11.

———. "Self-renewal of embryonic stem cells by a small molecule." *Proceedings of the National Academy of Sciences U.S.A.* 176 (2006): 17,266–71.

Shine, I., and S. Wrobel. *Thomas Hunt Morgan: Pioneer of Genetics.* Lexington: University of Kentucky Press, 1976.

Slack, J. M. W. *Stem Cells—A Very Short Introduction.* Oxford University Press, 2012.

Spence, J. R., et al. "Directed differentiation of human pluripotent stem cells into intestinal tissue in vitro." *Nature* 470 (2010): 105–9.

Stoick-Cooper, C. L., et al. "Distinct Wnt signaling pathways have opposing roles in appendage regeneration." *Development* 134 (2007): 479–89.

Suzuki, A., et al. "Liver repopulation by c-Met-positive stem/progenitor cells isolated from the developing rat liver." *Hepatogastroenterology* 51 (2004): 423–26.

Takahashi, K., and S. Yamanaka. "Induction of pluripotent stem cells from mouse embryonic and adult fibroblast cultures by defined factors." *Cell* 126 (2006): 663–76.

Takahashi, K., et al. "Induction of pluripotent stem cells from adult human fibroblasts by defined factors." *Cell* 131 (2007): 861–72.

Tanaka, Elly M. "Regeneration: If they can do it, why can't we?" *Cell* 113 (2003): 559–62.

Watt, F. M., and B. L. M. Hogan. "Out of Eden: Stem cells and their niches." *Science* 287 (2000): 1,427–30.

Wernig, M., et al. "In vitro reprogramming of fibroblasts into a pluripotent ES-cell-like state." *Nature* 448 (2007): 260–62.

Wilson, E. B. *The Cell in Development and Inheritance.* New York: Macmillan, 1896.

Wolpert, Lewis. *The Triumph of the Embryo.* New York: Oxford University Press, 1991.

Yamashita, Yukiko M., D. L. Jones, and M. T. Fuller. "Orientation of asymmetric stem cell division by the APC tumor suppressor and centrosome." *Science* 301 (2003): 1,547–50.

Yu, J., et al. "Induced pluripotent stem cell lines derived from human somatic cells." *Science.* Published online November 20, 2007 [DOI: 10.1126/science.1151526].

CHAPTER 3: CHALLENGERS OF ETHICS

Annas, G. J., and S. Elias. "Politics, morals and embryos." *Nature* 431 (2004): 19–20.

Barzun, Jacques. *From Dawn to Decadence: 500 Years of Western Cultural Life.* New York: HarperCollins, 2000.

Bush, George W. Executive Order: Expanding Approved Stem Cell Lines in Ethically Responsible Ways. June 20, 2007. http://georgewbush-whitehouse.archives.gov/news/releases/2007/06/20070620-6.html (accessed March 27, 2010).

Callahan, Daniel. *What Price Better Health? Hazards of the Research Imperative.* Berkeley: University of California Press, 2003.

Caplan, Arthur. "Is biomedical research too dangerous to pursue?" *Science* 303 (2004): 1,142.

Caplan, Arthur. "The stem cell hype machine." ScienceProgress. org, April 18, 2011. http://www.scienceprogress.org/2011/04/the-stem-cell-hype-machine/.

Carson, Benjamin S. Sr. Testimony before the President's Council on Bioethics, June 24, 2004.

Chew, Mabel. "Sex, Science & Society: Will sex survive to 2099?" *Medical Journal of Australia* 171 (1999): 659.

Chew, Matthew K., and Manfred D. Laubichler. "Essays on science and society: Natural enemies — metaphor or misconception?" *Science* 301 (2003): 52–53.

Cohen, Eric. "Bush's stem cell ruling: A Missouri Compromise." *Los Angeles Times,* August 12, 2001.

Edwards, R. G. "IVF and the history of stem cells." *Nature* 413 (2003): 349–51.

Executive Order 13505 of March 9, 2009, Removing Barriers to Responsible Scientific Research Involving Human Stem Cells. *Federal Register* Vol. 74, No. 46, March 11, 2009.

Executive Order 13521 of November 24, 2009, Establishing the Presidential Commission for the Study of Bioethical Issues. *Federal Register* Vol. 74, No. 228, November 30, 2009.

Fost, N. C. "Conception for donation." *Journal of the American Medical Association* 291 (2004): 2,126–27.

Fukuyama, Francis. *Our Posthuman Future: Consequences of the biotechnology revolution*. New York: Farrar, Straus and Giroux, 2002.

George, Robert. Testimony before the President's Council on Bioethics, January 16, 2003.

Gilbert, Scott F. "When does human life begin?" DevBio (a online companion to *Developmental Biology* by Scott F. Gilbert), April 2, 2002. http://8e.devbio.com/article.php?id=162 (accessed November 2, 2007).

Groopman, Jerome. "Science fiction." *New Yorker*, February 4, 2004.

Gurmankin, A. D., D. Sisti, and A. L. Caplan. "Embryo disposal practices in IVF clinics in the United States." *Politics & the Life Sciences* 22 (2003): 4–8.

Hoffman, D., et al. "Cryopreserved embryos in the United States and their availability for research." *Fertility and Sterility* 79 (2003): 1,063–69.

Johnson, Paul. *The Renaissance: A Short History*. New York: Modern Library, 2000.

Kass, Leon. Statement before the President's Council on Bioethics, February 13, 2002.

Kinsley, Michael. "Why Science Can't Save the GOP." *Time*, November 29, 2007.

Klimanskaya, I., et al. "Human embryonic stem cell lines derived from single blastomeres." *Nature* 444 (2006): 481–85.

Kono, T., et al. "Birth of parthenogenetic mice that can develop to adulthood." *Nature* 428 (2004): 860–64.

Krauthammer, Charles. "Stem Cell Vindication." *Washington Post,* November 30, 2007.

Leonardo daVinci. *The DaVinci Notebooks.* New York: Arcade Publishing, 2005.

Leshner, Alan I., and James A. Thomson. "Taking Exception: Standing in the Way of Stem Cell Research." *Washington Post,* December 3, 2007.

Levine, A. D. *Cloning: A Beginner's Guide.* Oxford, England: Oneworld Publications, 2007.

Lewontin, Richard C., and Richard Levins. "Stephen Jay Gould: What does it mean to be a radical?" *Monthly Review* 54 (6), November 2002. http://www.monthlyreview.org/1102lewontin.htm (accessed November 2, 2007).

Leydesdorff, Loet. "Meaning, Metaphors, and Translation at the Interfaces of Science: Mapping the Case of 'Stem-Cell Research.'" Science & Technology Dynamics, University of Amsterdam. Amsterdam School of Communications Research (ASCoR), 2003.

Machaty, Z., and R. S. Prather. "Complete oocyte activation using an oocyte-modifying agent and a reducing agent." Assignee: The

Curators of the University of Missouri. U.S. Patent No. 6,211,429, filed June 18, 1998, issued April 3, 2001.

Martin, M. J., et al. "Human embryonic stem cells express an immunogenic nonhuman sialic acid." *Nature Medicine* 11 (2005): 228–32.

Nicholl, Charles. *Leonardo da Vinci: Flights of the Mind*. New York: Viking, 2004.

Pollack, A. "Debate on human cloning turns to patents." *New York Times*, May 17, 2002.

President's Council on Bioethics. "Human cloning and human dignity: An ethical inquiry." July, 2002. http://bioethics. georgetown.edu/pcbe/reports/cloningreport/ (accessed March 26, 2011).

———. "Being Human: Readings from the President's Council on Bioethics." December 2003. http://bioethics.georgetown.edu/ pcbe/bookshelf/ (accessed March 26, 2011).

———. "Beyond Therapy: Biotechnology and the Pursuit of Happiness." October 2003. http://bioethics.georgetown.edu/pcbe/ reports/beyondtherapy/ (accessed March 26, 2011).

———. "Monitoring Stem Cell Research." January, 2004. http:// bioethics.georgetown.edu/pcbe/reports/stemcell/ (accessed March 26, 2011).

———. Reproduction and Responsibility: The Regulation of New Biotechnologies." March 2004. http://bioethics.georgetown.edu/

pcbe/reports/reproductionandresponsibility/ (accessed March 26, 2011).

Saletan, William. "Oy Vitae: Jews vs. Catholics in the stem cell debate." *Slate,* March 12, 2005. http://slate.com/id/2114733/ (accessed November 2, 2007).

Silver, Lee. "Stem-Cell Semantics" (video interview with reporter Carl Zimmer). *New York Times.* http://video.on.nytimes.com/ (accessed March 26, 2011).

Stevens, M. L. T. *Bioethics in America: Origins and Cultural Politics.* Baltimore: Johns Hopkins University Press, 2000.

Stock, G., and F. Fukuyama. "The Genetic Future Is Now: Redesigning Humans vs. Regulating Science." Book Forum, Cato Institute, March 15, 2002. http:// www.cato.org/events/020315bf.html (accessed November 2, 2007).

Vogelstein, B., B. Alberts, and K. Shine. "Please don't call it cloning." *Science* 295 (2002): 1,237.

Walters, L. "Human embryonic stem cell research: An intercultural perspective." *Kennedy Institute of Ethics Journal* 14 (2004): 3–38.

Warnock, Mary. *A Question of Life: The Warnock Report on Human Fertilisation and Embryology.* London: Basil Blackwell, 1985.

Weintraub, A. "Repairing the engines of life." *BusinessWeek,* May 24, 2004.

Weiss, R. "Funding bill gets clause on embryo patents." *Washington Post*, November 17, 2003.

World Stem Cell Policy Map, University of Minnesota. http://www.mbbnet.umn.edu/scmap.html (accessed March 26, 2011).

Zhao, X., et al. "iPS cells produce viable mice through tetraploid complementation." *Nature* 461(2009): 86–90.

CHAPTER 4: BAROMETERS OF POLITICS

Brustle v. Greenpeace eV, European Court of Justice Case C-34/10. Opinion of Advocate General Yves Bot, March 10, 2011.

Bryan, William Jennings. Text of "Cross of Gold" speech at 1896 Democratic National Convention. Public Broadcasting System. http://www.pbs.org/wgbh/amex/1900/filmmore/reference/primary/crossofgold.html (accessed November 2, 2007).

Chakrabarty, Ananda M. "Crossing species boundaries and making human-nonhuman hybrids: Moral and legal ramifications." *American Journal of Bioethics* 3 (2003): 20–21.

Commonwealth of Massachusetts et al. v. Environmental Protection Agency et al. U.S. Supreme Court Docket No. 05-1120 (2006). Oral arguments involving U.S. Supreme Court Justice Antonin Scalia.

Diamond v. Chakrabarty. U.S. Supreme Court Docket No. 79–136. 447 U.S. 303 (1980).

Dickey-Wicker Amendment. The Balanced Budget Downpayment Act, I. Public Law 104-99, signed into law on Jan. 26, 1996. Available at: http://www.gpo.gov/fdsys/pkg/PLAW-104publ99/html/PLAW-104publ99.htm (accessed March 14, 2011).

Economist. "A nation apart." November 6, 2003.

Hammer v. Dagenhart. U.S. Supreme Court Docket No. 247 U.S. 251 (1918), Justice Oliver Wendell Holmes Jr. dissenting. Holmes, Oliver Wendell Jr. 1881 *The Common Law.* Boston: Little, Brown, 1881.

Huxley, Aldous. *Brave New World* (with new foreword by the author). New York: HarperPerennial, 1946.

Kerry, John. Text of acceptance speech at 2004 Democratic National Convention. Online NewsHour, Public Broadcasting System. http://www.pbs.org/newshour/vote2004/demconvention/speeches/kerry.html (accessed November 2, 2007).

Korobkin, Russell and Stephen R. Munzer. *Stem Cell Century: Law and Policy for a Breakthrough Technology.* New Haven, Conn.: Yale University Press, 2007.

Levine, A. D. "Policy uncertainty and the conduct of stem cell research." *Cell Stem Cell* 8 (2011): 132–135.

Micklethwait, John, and Adrian Wooldridge. *The Right Nation: Conservative Power in America.* New York: Penguin, 2005.

Nisbet, Matthew C., Dominique Brossard, and Adrianne Kroepsch. "Framing Science: The stem cell controversy in an age of press/

politics." *Harvard International Journal of Press/Politics* 8 (2003): 3,670.

The Pew Research Center for the People and the Press. "Public Makes Distinctions on Genetic Research," April 9, 2002. http://people-press.org/reports/display.php3?ReportID=152 (accessed November 2, 2007).

Reagan, Ron. Text of speech at 2004 Democratic National Convention. Online NewsHour, Public Broadcasting System. http://www.pbs.org/newshour/vote2004/demconvention/speeches/reagan.htm (accessed March 26, 2011).

Rosen, Jeffrey. "Roberts v. The Future." *New York Times Magazine*, August 28, 2005.

Sachedina, Abdulaziz. "Islamic Perspectives on Cloning." Testimony before the National Bioethics Advisory Commission, March 14, 1997.

Sherley et al. v. Sebelius et al. U.S. Court of Appeals for the District of Columbia. Case: *10–5287*, Document: 1272139. Filed: October 18, 2010.

Tocqueville, Alexis de. *Democracy in America.* Translated and edited by Harvey C. Mansfield and Delba Winthrop. University of Chicago Press, 2000 (originally published 1835 and 1840).

Twain, Mark (Samuel L. Clemens). "Corn-pone Opinions" (1901), in Joyce Carol Oates and Robert Atwan, eds., *The Best American Essays of the Century*. Boston: Houghton Mifflin, 2001.

United Nations. "Ad Hoc Committee on an International Convention against the Reproductive Cloning of Human Beings." 2001. http://www.un.org/law/cloning/ (accessed November 2, 2007).

United Nations. "General Assembly adopts United Nations declaration on human cloning by vote of 84–34–37." Press release GA/10333, March 8, 2005.

United States Congress. "Stem Cell Research Enhancement Act of 2007." http://www.govtrack.us/congress/bill.xpd?bill=s110-5 (accessed November 2, 2007).

———. Stem Cell Research Enhancement Act of 2005 (Castle-DeGette). http://www.govtrack.us/congress/bill.xpd?bill=h109-810 (accessed November 2, 2007).

———. Stem Cell Therapeutic and Research Act of 2005 signed into law December 2005 establishing a national human cord blood bank. http://www.govtrack.us/congress/bill.xpd?bill=h109-2520 (accessed November 2, 2007).

United States Department of State. "Views of the United States on Cloning." September 23, 2002. http://www.state.gov/s/l/38722.htm (accessed November 2, 2007).

Winthrop, Robert C. *Life and Letters of John Winthrop, Governor of the Massachusetts Bay Company at Their Emigration to New England.* Boston: Little, Brown, 1869.

CHAPTER 5: OBJECTS OF COMPETITION

Almond, G. A., R. S. Appleby, and E. Sivan. *Strong Religion: The Rise of Fundamentalisms Around the World.* University of Chicago Press, 2002.

Arnold, W. "Singapore goes for biotech." *New York Times,* August 26, 2003.

Belluck, P. "Massachusetts proposes stem cell research grants." *New York Times,* May 9, 2007.

Broad, William J., and James Glanz. "Does science matter?" *New York Times,* November 7, 2003.

Bush, George W. "American Competitiveness Initiative." http:// georgewbush-whitehouse.archives.gov/stateoftheunion/2006/ aci/(accessed March 26, 2011).

————. "President Discusses Stem Cell Research." August 9, 2001. http://georgewbush-whitehouse.archives.gov/news/releases/ 2001/08/20010809-2.html (accessed March 26, 2011).

————. State of the Union speech. January 31, 2006. http:// georgewbush-whitehouse.archives.gov/stateoftheunion/2006/ (accessed March 27, 2011).

California Senate Bill No. 471, The *California* Stem Cell and Biotechnology Education and Workforce Development Act of 2009, signed into law October 11, 2009.

Diamond, Jared. *Guns, Germs and Steel: The Fate of Human Societies*. New York: W. W. Norton, 1997.

Duga, J., and T. Studt. "Global R&D report: Globalization alters traditional R&D roles." *R&D Magazine*, September 2006.

Florida, Richard. *The Rise of the Creative Class: And How It's Transforming Work, Leisure and Everyday Life*. New York: Basic Books, 2002.

Foer, Franklin. "The Joy of Federalism." *New York Times*, March 6, 2005.

Friedman, Thomas. "Losing Our Edge?" *New York Times*, April 22, 2004.

Golden, Frederic. "StemWinder." *Time*, August 20, 2001.

Harris Poll No. 57. (2003) "Scientists, Firemen, Doctors, Teachers and Nurses Top List as 'Most Prestigious Occupations.'" http://www.harrisinteractive.com/. . ./Harris-Interactive-Poll-Research-Pres-Occupations-2007-08.pdf (accessed March 27, 2011).

Hwang, Woo-suk, et al. "Evidence of a pluripotent human embryonic stem cell line derived from a cloned blastocyst." *Science* 303 (2004): 1,669–74.

———. "Patient-specific embryonic stem cells derived from human SCNT blastocysts." *Science* 308 (2005): 1,777–83.

International Society for Stem Cell Research (ISSCR). ISSCR's Global Membership. http://www.isscr.org (accessed March 27, 2011).

International Stem Cell Forum (ISCF). The ISCI Project. http://www.stem-cell-forum.net/ (accessed June 3, 2011).

Johnson, Paul. *The Renaissance: A Short History*. New York: Modern Library, 2000.

Juvenile Diabetes Research Foundation (JDRF). Fact Sheets: Research Funding Facts. http://www.jdrf.org/index.cfm?page_id=101006 (accessed November 2, 2007).

————. JDRF-Funded Research. http://onlineapps.jdfcure.org/ *AbstractSearchEngine.cfm* (accessed November 2, 2007).

Kafatos, Fotis C., and Thomas Eisner. "Unification in the century of biology." *Science* 303 (2004): 1,257.

Karmali, R. N., N. M. Jones, and A. D. Levine. "Tracking and assessing the rise of state-funded stem cell research." *Nature Biotechnology* 28 (2011): 1,246–48.

Kim, K. S., et al. "Self-renewal induced efficiently, safely, and effective therapeutically with one regulatable gene in a human somatic progenitor cell." *Proceedings of the National Academy of Sciences U.S.A.* Published online March 4, 2011 [DOI:10.1073/pnas.1019743108].

Leonardo daVinci. *The DaVinci Notebooks.* New York: Arcade Publishing, 2005.

Livingstone, David N. *Putting Science in Its Place: Geographies of Scientific Knowledge.* University of Chicago Press, 2003.

McMahon, D. S., et al. "Cultivating regenerative medicine innovation in China." *Regenerative Medicine* 5 (2010): 35–43.

Miller, Greg. "CIRM: The good, the bad, and the ugly." *Science* 330 (2010): 1,742–43.

National Academy of Sciences. *Rising Above the Gathering Storm: Energizing and Employing America for a Brighter Economic Future.* Washington, D.C.: National Academies Press, 2005.

National Institutes of Health (NIH). Estimates of Funding for Various Diseases, Conditions, Research Areas. http://report.nih.gov/rcdc/categories/ (accessed March 27, 2011).

National Science Foundation, Division of Science Resources Statistics, 2003. Federally Funded R&D for National Defense and Civilian Functions: fiscal years 1955–2003. www.nsf.gov/statistics/nsf02330/pdf/tab4.pdf (accessed November 2, 2007).

National Science Foundation. Science and Engineering Indicators 2010. http://www.nsf.gov/statistics/seind10/ (accessed April 29, 2011).

Normile, D. "Singapore-Hopkins partnership ends in a volley of fault-finding." *Science,* August 4, 2006.

Office of the California Secretary of State. Proposition 71 (2004). Available via the California Institute of Regenerative Medicine. http://www.cirm.ca.gov/pdf/prop71.pdf (accessed November 2, 2007).

Pilcher, H. R. "Britain's Stem-Cell Store Opens: Cell bank will serve researchers worldwide." *Nature*, May 20, 2003.

Porter, Michael E. *The Competitive Advantage of Nations*. New York: Free Press, 1990.

Qiu, Jane. "China Spinal Cord Injury Network: changes from within" *The Lancet Neurology* 8 (2009): 606–607.

ScanBalt BioRegion. http://www.scanbalt.org (accessed November 2, 2007).

Seoul National University Investigation Committee. "Summary of the final report on Professor Woo-suk Hwang's research allegations." *New York Times*, January 6, 2006.

Sipp, Douglas. "Stem cell research in Asia: A critical review." *Journal of Cellular Biochemistry* 107 (2009): 853–856.

———. "Stem cells and regenerative medicine on the Asian horizon: an economic, industry and social perspective." *Regenerative Medicine* 6 (2009): 911–917.

Spar, Debora L. *The Baby Business: How Money, Science, and Politics Drive the Commerce of Conception*. Cambridge, Mass.: Harvard Business School Press, 2006.

State of New Jersey. Senate Bill No. 1909. Synopsis: Permits human stem cell research in New Jersey. Introduced September 30, 2002. http://www.njleg.state.nj.us/2002/Bills/S2000/1909_I1.HTM (accessed November 2, 2007).

Stem Cell Network—Asia-Pacific (SNAP). http://www.asiapacific-stemcells.org (accessed March 14, 2011).

Summers, Lawrence. Presidency of Harvard University inaugural address, October 12, 2001. http://president.harvard.edu/speeches/summers_inauguration/summers.php (accessed March 27, 2011).

U.S. Congress. America COMPETES Act. Public Law 110-69, signed into law on August 9, 2007. http://www.govtrack.us/congress/bill.xpd?bill=h110-2272 (accessed December 5, 2007).

Wild, S. et al. "Global prevalence of diabetes." *Diabetes Care* 27 (2004): 1047–53.

Wilmut, et al. "Viable offspring derived from fetal and adult mammalian cells." *Nature* 385 (1997): 810–3.

Wisconsin Alumni Research Foundation (WARF). Stem Cell Patents. Public Patent Foundation, New York. http://www.pubpat.org/warfstemcell.htm (accessed November 2, 2007).

World Health Organization. "Launch of 'Diabetes Action Now': New estimate of more than three million diabetes-related deaths every year." May 5, 2004. http://www.who.int/mediacentre/news/releases/2004/pr31/en/ (accessed November 2, 2007).

Yergin, Daniel, and Joseph Stanislaw. *The Commanding Heights: The Battle between Government and the Marketplace That Is Remaking the Modern World.* New York: Simon & Schuster, 1998.

Young, Robin R. *Stem Cell Analysis and Market Forecasts 2006–2016.* Wayne, Pa.: RRY Publications, 2006.

Yuan, Robert. "Korea lays groundwork for future success." *Genetic Engineering News,* Jan. 1, 2009.

CHAPTER 6: HARBINGERS OF DESTRUCTION

Ball, P. "Starting from scratch." *Nature* 431 (2004): 624–26.

"BARDA Funds Medical Countermeasure Innovation." Biomedical Advanced Research and Development Authority (BARDA). U.S. Department of Health and Human Services. http://www.phe.gov/Preparedness/mcm/Pages/innovation.aspx (accessed March 27, 2010).

BioBricks Foundation. http://www.biobricks.org/ (accessed November 2, 2007).

Blomfield, A. "Russia bans specimen exports." *Telegraph* (U.K.), June 2, 2007.

Callahan, Gerald N. *Faith, Madness and Spontaneous Human Combustion: What Immunology Can Teach Us about Self-Perception.* New York: St. Martin's Press, 2002.

Callahan, Michael. "Engineering Bio-Terror Agents: Lessons from the Offensive U.S. and Russian Biological Weapons Programs." Hearing before the Subcommittee on Prevention of Nuclear and Biological Attack of the Committee on Homeland Security, U.S. House of Representatives, July 13, 2005. http://www.fas.org/irp/congress/2005_hr/bioterror.html (accessed March 27, 2011).

Carlson, Rob. "The pace and proliferation of biological technologies." *Biosecurity and Bioterrorism: Biodefense Strategy, Practice, and Science* 1 (2003): 1–12.

Clark, Kenneth. *Leonardo daVinci: An Account of His Development as an Artist.* New York: Macmillan, 1939.

Cliatt, C. "Princeton scientists explore the next frontier of stem cell research." *Princeton Weekly Bulletin,* June 19, 2006.

Commission on Prevention/WMDs. *World at Risk: The Report of the Commission on the Prevention of WMD Proliferation and Terrorism.* Bob Graham, Chairman; Jim Talent, Vice-Chairman, eds. New York: Vintage Books, 2008.

Crichton, Michael. *The Andromeda Strain.* New York: Dell, 1969.

Defense Advanced Research Projects Agency (DARPA). Fundamental Laws of Biology. http://www.darpa.mil/ . . . /Fundamental_Laws_of_Biology_(FUNBIO).aspx (accessed March 27, 2011).

Defense Advanced Research Projects Agency (DARPA), 2003. Fiscal Year (FY) 2004 / FY 2005 Biennial Budget Estimates. "Develop an integrated in vitro human immune system, capable of supporting rapid and cost effective vaccine development and testing through the establishment of tools necessary for in vitro fabrication of three dimensional tissue constructs, bioscaffolds and bioreactors," 131.

Economist. "Science and the Bush administration: Cheating nature?" April 7, 2004.

Fauci, Anthony S. "BioShield: Countering the Bioterrorist Threat." Testimony before the U.S. House of Representatives Select Committee on Homeland Security, May 15, 2003.

Fauci, A. S., R. Hirschberg, and J. La Montagne. "Biomedical research: An integral component of national security." *New England Journal of Medicine* 350 (2004): 2,120–22.

FedBizOpps (FBO) Daily Issue of January 19, 2002. BAA-0142, Addendum 8, Special Focus Area: Engineered Tissue Constructs (ETC). Contracting office: Defense Advanced Research Projects Agency. Notice date: January 17, 2002.

Galic, Z., et al. "T lineage differentiation from human embryonic stem cells." *Proceedings of the National Academy of Sciences U.S.A.* 103 (2006): 11,742–47.

Gerberding, Julie. "The State of the CDC: Fiscal Year 2004: Protecting Health for Life." Remarks at the National Press Club Conference, February 22, 2005. http://www.cdc.gov/media/pressrel/r050222b.htm (accessed March 27, 2011).

Gould, Stephen Jay. "Essays on science and society: The great asymmetry." *Science* 279 (1998): 812–13.

———. *The Structure of Evolutionary Theory.* Cambridge, Mass.: Harvard University Press, 2002.

Hoffman, David E. "Going Viral: The Pentagon takes on a new enemy: swine flu." *New Yorker,* January 31, 2011.

Homeland Security Advanced Research Projects Agency (HSARPA). http://www.dhs.gov/files/grants/gc_1247254578009.shtm (accessed March 27, 2011).

Kotov, N. A., et al. "Inverted colloidal crystals as three-dimensional cell scaffolds." *Langmuir* 20 (2004): 7,887–92.

Langman, Rodney E. *The Immune System: Evolutionary Principles Guide Our Understanding of This Complex Biological Defense System.* New York: Academic Press, 1989.

Leonardo daVinci. *The DaVinci Notebooks,* New York: Arcade Publishing, 2005.

Licklider, J. C. R. Topics for Discussion at the Forthcoming Meeting. Memorandum To: Members and Affiliates of the Intergalactic Computer Network. Advanced Research Projects Agency,

April 23, 1963. http://www.kurzweilai.net/memorandum-for-members-and-affiliates-of-the-intergalactic-computer-network (accessed March 27, 2011).

Lucretius. *On the Nature of Things (De rerum natura)*. Available at Website of Project Gutenberg: http://www.gutenberg.org/etext/785 (accessed November 2, 2007).

May, Mike. "Stem cells serve as new platform for biodefense preparedness." *Nature Medicine* 16 (2010): 835–836.

Musallam, A. *From Secularism to Jihad: Sayyid Qutb and the Foundations of Radical Islamism*. Westport, Conn: Greenwood Publishing, 2005.

National Science Advisory Board for Biosecurity (NSABB). http://oba.od.nih.gov/biosecurity/about_nsabb.html (accessed June 3, 2011).

Office of Transnational Issues. "The Darker Bioweapons Future." U.S. Central Intelligence Agency, Directorate of Intelligence, November 3, 2003.

openDOOR (MIT Alumni Association). "Bioengineering and Beyond," October 2001. https://alum.mit.edu/ne/opendoor/index.html (accessed November 2, 2007).

Orlando Regional Chamber of Commerce. "Breaking Out: Innovative companies are diversifying the region's economy while making a lasting impact on their industries." *First Monday* 10 (2006): 14–18.

Pandemic and All-Hazards Preparedness Act. Public Law No. 109–417, signed into law December 19, 2006. http://www.govtrack.us/congress/bill.xpd?bill=s109-3678 (accessed November 2, 2007).

Priest, Dana, and William M. Arkin. "Top secret America: A Washington Post investigation." *Washington Post*, July–December, 2010.

Rhodes, Richard. *The Making of the Atomic Bomb.* New York: Simon & Schuster, 1986.

Richardson, L. "Buying Biosafety: Is the price right?" *New England Journal of Medicine* 350 (2004): 2,121–23.

Slukvin, I. I., et al. "Directed differentiation of human embryonic stem cells into functional dendritic cells through the myeloid pathway." *Journal of Immunology* 176 (2006): 2,924–32.

————. Method of forming dendritic cells from embryonic stem cells. United States Patent Application 20060275901, filed December 7, 2006.

Tether, Tony. Statement by Dr. Tony Tether, Director, Defense Advanced Research Projects Agency, submitted to the Subcommittee on Terrorism, Unconventional Threats and Capabilities, House Armed Services Committee, U.S. House of Representatives, March 27, 2003. http://www.darpa.mil/WorkArea/DownloadAsset.aspx?id=1778 (accessed November 2, 2011).

U.S. Congress. The Project BioShield Act of 2004. Public Law 108-276, signed into law on July 21, 2004. http://www.govtrack.us/congress/bill.xpd?bill=s108-15 (accessed March 27, 2011).

Vodyanik, M. A., J. A. Thomson, and I. I. Slukvin. "Leukosialin (CD43) defines hematopoietic progenitors in human embryonic stem cell differentiation cultures." *Blood* 108 (2006): 2,095–2,105.

Warren, William, et al. Automatable artificial immune system (AIS). United States Patent Application 20060270029, filed November 30, 2006.

Weiss, R., and I. Lemischka. Programmed differentiation of mouse embryonic stem cells using artificial signaling pathways. iGEM 2006 (The international Genetically Engineered Machine competition), Massachusetts Institute of Technology, 2006.

White House. "Biodefense for the 21st Century," April 28, 2004. http://www.fas.org/irp/offdocs/nspd/hspd-10.html (accessed March 27, 2011).

Wolfe, N. D., C. Panosian Dunavan, and J. Diamond. "Origins of major human infectious diseases." *Nature* 447 (2007): 279–83.

Wu, J. H. David. Ex vivo generation of functional leukemia cells in a three-dimensional bioreactor. United States Patent 7,087,431, filed March 1, 2001, issued August 8, 2006.

Yachie, N., et al. "Alignment-based approach for durable data storage into living organisms." *Biotechnology Progress* 23 (2007): 501–5.

Yergin, Daniel. "Fighting the globalization flu." *The Globalist*, May 29, 2003.

EPILOGUE: BEYOND THE DARKNESS

Leonardo daVinci. *The DaVinci Notebooks.* New York: Arcade Publishing, 2005.

Nicholl, Charles. *Leonardo da Vinci: Flights of the Mind.* New York: Viking, 2004.

Nuland, Sherwin. *Leonardo daVinci.* New York: Penguin, 2001.

INDEX